Markovian
Decision
Processes

Modern Analytic *and* Computational Methods *in* Science *and* Mathematics

A GROUP OF MONOGRAPHS
AND ADVANCED TEXTBOOKS

Richard Bellman, EDITOR
University of Southern California

Published:

In Preparation

Markovian Decision Processes

Hisashi Mine

and

Shunji Osaki

Department of Applied Mathematics and Physics
Faculty of Engineering, Kyoto University
Kyoto, Japan

American Elsevier
Publishing Company, Inc.
NEW YORK · 1970

AMERICAN ELSEVIER PUBLISHING COMPANY, INC.
52 Vanderbilt Avenue, New York, N.Y. 10017

ELSEVIER PUBLISHING COMPANY, LTD.
Barking, Essex, England

ELSEVIER PUBLISHING COMPANY
335 Jan Van Galenstraat, P.O. Box 211
Amsterdam, The Netherlands

International Standard Book Number 0–444–00079–8

Library of Congress Card Number 70–116709

Printed in Scotland

Contents

Preface

A Markovian decision process is a sequential decision process on a discrete-time Markov chain. Since 1960 many contributions to Markovian decision processes have been made, and their extensions and applications have also been developed. However, no book has previously been published on standard Markovian decision processes. This work is intendedto describe Markovian decision processes by emphasizing algorithms.

A Markovian decision process is a simple mathematical model, and therefore possesses elegant results. General structures and properties included in Markovian decision processes are of great interest not only to operations analysts and applied mathematicians, but also to statisticians, economists, and engineers.

This book is mainly concerned with the mathematical theory and algorithms of Markovian decision processes. Furthermore, some extended models, for example, semi-Markovian decision processes, general sequential decision processes and stochastic games, are also discussed. These models are closely related to Markovian decision processes.

We do not intend to describe all branches of Markovian decision processes but rather to develop some of the most important types. Further details concerning special topics and applications will be found in the bibliography. To understand this book, only a knowledge of elementary Markov chain theory and linear programming theory are assumed.

The authors wish specifically to thank Professor Richard Bellman for his encouragement and for his acceptance of this book in his series.

It would have been impossible to complete this work without the assistance we have received from certain individuals and organizations. We are indebted to the editors of *Annals of Mathematical Statistics*, *SIAM*, *Management Science*, and *Operations Research* for permission to quote material that has previously been published in these journals. We also wish to thank Academic Press for permission to quote our papers. Special mention should be made to M.I.T. Press, which granted permission to reproduce some numerical examples from "Dynamic Programming and Markov Processes" by R. A. Howard. Thanks are also due to the members of Professor Mine's laboratory of the Department of Applied Mathematics and Physics of the Kyoto University. Finally we wish to thank Miss Noriko Tsuji for typing the manuscript.

HISASHI MINE
SHUNJI OSAKI

Kyoto, Japan
October, 1969

Chapter 1

Introduction

A Markovian decision process is a stochastic sequential process. Consider a system that can be described by a discrete-time Markov chain, where, furthermore, the decisions of each epoch and the returns are associated with each state we observe.

As a simple example of Markovian decision processes, consider a machine maintenance problem. A machine can be operated synchronously, say, once an hour. At each period there are two states; one is operating (state 1), and the other is a condition of failure (state 2). If the machine fails, the machine can be restored to perfect functioning by repair. At any period, if the machine is running, we earn the return of $3.00 per period; the probability of being in state 1 at the next step is 0.7 and the probability of moving to state 2 is 0.3. If the machine is in failure condition, we have two actions to repair the failed machine; one is a rapid repair that requires a cost of $2.00 (that is, a return of − $2.00), with the probability of moving to state 1 of 0.6; another is a usual repair that yields the cost $1.00 (that is, a return of − $1.00), with the probability of moving to state 1 of 0.4. For the model considered, there are two alternatives available in each state. The state transition diagrams of two such alternatives are shown in Fig. 1.1.

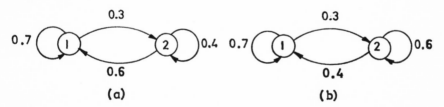

Fig. 1.1. Transition diagrams for the machine maintenance problem. (a) Rapid repair. (b) Usual repair.

Under these situations, what strategies may we choose to maximize the total expected return starting in some state? We now consider a time-planning horizon. If a time-planning horizon is finite, then we call this problem a "finite-horizon problem." A finite-horizon problem can be directly formulated by dynamic programming, and an optimal strategy that is a sequence of decisions for each state can be obtained by means of a recurrence relation. If, however, a time-planning horizon is infinite, which we call an "infinite-horizon problem," we cannot apply either the dynamic programming approach or the direct enumeration approach because the strategies considered are infinite. Furthermore, the total expected return will usually be divergent. We consider then two approaches for the problem over an infinite horizon. One is an introduction of a discount factor; a problem of this type is called a "discounted problem" or a "problem with discounting." A unit return at the n period is worth only β^n if we define a discount factor β $(0 \leqslant \beta < 1)$, and the total expected return is always convergent. For a discounted problem, we may apply the total expected return as our objective function. Or we may consider a problem with no discounting. For a nondiscounted problem, we may apply the average return per period in the steady state as our objective function if the total expected return is divergent. In Chapter 3 we consider a case where the total expected return is convergent for a nondiscounted problem.

Our main concerns are to find a strategy for maximizing the given objective function and to determine its maximum value. Such processes are called Markovian decision processes.

Markovian decision processes are based on discrete-time Markov chains. We can extend a discrete-time process to a continuous-time process. Let us consider an example similar to the preceding one. In the preceding example, we considered that the machine operates in discrete time. We can also consider a machine that can be operated continuously in time. The distribution of the running time is $F_{12}(t)$ $(t \geqslant 0)$, while the two distributions of completing a repair are $F_{21}^k(t)$ $(t \geqslant 0)$, $k = 1$ and 2, respectively, where k denotes any action in state 2. Furthermore, we may consider returns that depend on the duration of the elapsed time in any state. Such a process is based on a semi-Markov process or a Markov renewal process; we call it a semi-Markovian decision process or a Markov renewal programming process. As a planning horizon we can consider either a finite-horizon or an infinite-horizon problem. Here we consider only an infinite-horizon problem for semi-Markovian decision processes.

Another extension can also be considered. The process may be discrete in time, with the state space and the action space any Borel sets. Such a process is called a generalized Markovian decision process. We consider here only a discounted case.

Markovian decision processes with discounting are considered in Chapter 2. Problems over an infinite horizon are formulated by means of the policy iteration algorithm and a linear programming algorithm. In addition, the close relation between the two approaches is described, and some numerical examples presented. Markovian decision processes with no discounting are considered in Chapters 3 and 4. A completely ergodic process and a terminating process are described in Chapter 3. A general process is treated in Chapter 4. Markovian decision processes over a finite horizon are described in Chapter 5 from the viewpoint of dynamic programming.

Semi-Markovian decision processes are discussed in Chapter 6. A semi-Markovian process is defined and its properties are presented. Several problems we have discussed in Markovian decision processes are also formulated as linear programming problems. These linear programming problems use corresponding policy iteration algorithms.

Generalized Markovian decision processes are considered in Chapter 7. For the discounted process, some existence theorems for optimal plans are presented. General decision processes are formulated in Chapter 8. The contraction (or the N-stage contraction) and monotonicity properties in a metric space are discussed, and then optimization schemes for finding or approximating optimal strategies are presented.

Stochastic games, both important and interesting, are considered in the appendix. If one of two players is a dummy, the stochastic game reduces to a Markovian decision process.

Throughout this book comments are added at the end of each chapter presenting the source of materials and further related works. The conclusion in Chapter 9 presents some history and further works on Markovian decision processes, as well as on applications. A bibliography is given at the end of this book.

Chapter 2

Markovian Decision Processes with Discounting

2.1. Introduction

Consider a system whose *state space* has finitely many states, as we have described in the preceding chapter. Let a state space S be a set of states labeled by the integers $i = 1, 2, \ldots, N$. That is, $S = \{1, 2, \ldots, N\}$. For each $i \in S$, we have a set K_i of finite *actions* (or *alternatives*) labeled by the integers $k = 1, 2, \ldots, K_i$.

The *policy space* is denoted by the Cartesian product of each action set, that is, $K = K_1 \times K_2 \times \cdots \times K_N$. Next, consider a sequential decision problem, that is, periodically observe one of the states at time $n = 0, 1, 2, \ldots$, and make an action at each time.

When the system is in state $i \in S$ and we make an action $k \in K_i$, two things happen: (1) We obtain the *return* r_i^k. (2) The system obeys the probability law p_{ij}^k ($j \in S$) at the next time, where p_{ij}^k is the *transition probability* that the system is in state j at the next time, given that the system is in state i at that time and an action k is made. Here we assume that the return r_i^k is bounded for all $i \in S$ and $k \in K_i$. It is clear from the finiteness of state space that

$$\sum_{j \in S} p_{ij}^k = 1, \qquad p_{ij}^k \geqslant 0 \qquad \text{for} \quad i, j \in S, \quad k \in K_i \tag{2.1}$$

In this chapter, we consider a discounted process. Let β $(0 \leqslant \beta < 1)$ be a *discount factor*. That is, the unit return becomes β^n after n times (for example, n days). The discount factor is considered as the reciprocal of 1 plus the interest rate. The introduction of the discount factor is to avoid mathematically the divergence of the total expected return.

We also give an initial distribution

$$a = (a_1, a_2, \ldots, a_N) \tag{2.2}$$

where

$$\sum_{i \in S} a_i = 1, \qquad a_i \geqslant 0 \qquad \text{for} \quad i \in S \tag{2.3}$$

The system is then a nonstationary Markov chain with returns. Our problem is to find *strategies* that maximize the discounted total expected return over a finite-time or an infinite-time horizon, where a strategy is a sequence of decisions in each time and each state.

In this chapter we shall focus our attention on a discounted decision process over an infinite-time horizon. We consider a maximization problem (if we consider a minimization problem, we may change the sign of the returns).

2.2. Policy Iteration Algorithm

In this section we shall give the *policy iteration algorithm* for a discounted Markovian decision process. The policy iteration algorithm was first given by Howard [63], and thus is sometimes called Howard's policy iteration algorithm. The policy iteration algorithm is closely related to linear programming, which will be discussed later. The discussion in this section forms a basis for the discussion of the general decision processes in Chapter 8.

Since we consider a sequential decision process, we have complete information on all states and times. Let F be a set of functions from the state space S to the policy space K. Since S and its associated K are both finite sets, F is a finite set. Let f or g be a function in F. Then a *strategy* π is defined by a sequence $\{f_n, n = 1, 2, \ldots\}$. Hence, we may write a strategy

$$\pi = (f_1, f_2, \ldots, f_n, \ldots),$$

where f_n is the decision vector for each state at time n; that is, $f_n(i)$, the ith element of f_n, is an action of state $i \in S$ at time n.

A strategy $(g, f_1, f_2, \ldots)$ is denoted by (g, π), where $g \in F$ and $\pi = (f_1, f_2, \ldots)$. A strategy $(f, f, \ldots, f, \ldots)$, denoted by f^∞, where $f \in F$, is called a *stationary strategy*. That is, a stationary strategy f^∞ is one that is independent of time n. A strategy

$$(\overbrace{g, g, \ldots, g}^{n}, f_1, f_2, \ldots)$$

is denoted by (g^n, π), where $g \in F$ and $\pi = (f_1, f_2, \ldots)$.

For any strategy π, we have a nonstationary Markov chain. Thus we write the n-step transition probability matrix as

$$P_n(\pi) = P(f_1)P(f_2) \cdots P(f_n) \qquad \text{for} \quad n = 1, 2, \ldots \qquad (2.4)$$

where $P(f_n)$ is the $N \times N$ transition matrix whose $i-j$th element is p_{ij}^k, $k = f_n(i) \in K_i$. For $n = 0$, we define $P_0(\pi) = I$ (the $N \times N$ identity matrix). For any $f \in F$, we may write the $N \times 1$ return vector $r(f)$ whose ith element is r_i^k, $k = f(i) \in K_i$. Under the notation defined above, we have the $N \times 1$ discounted total expected return vector starting in each state $i \in S$:

$$V_\beta(\pi) = \sum_{n=0}^{\infty} \beta^n P_n(\pi) r(f_{n+1}) \qquad (2.5)$$

To show the finiteness of this vector, setting $r^U = \max_{i,k} r_i^k$ and $r^L = \min_{i,k} r_i^k$, we have

$$\frac{r^L}{(1-\beta)}\mathbf{1} \leqslant V_\beta(\pi) \leqslant \frac{r^U}{(1-\beta)}\mathbf{1} \tag{2.6}$$

where $\mathbf{1}$ is the $N \times 1$ vector with all elements unity.

We have

$$\begin{aligned} V_\beta(\pi) &= \sum_{n=0}^\infty \beta^n P_n(\pi) r(f_{n+1}) \\ &= r(f_1) + \sum_{n=1}^\infty \beta^n P_n(\pi) r(f_{n+1}) \\ &= r(f_1) + \beta P(f_1) \sum_{n=0}^\infty \beta^n P_n(T\pi) r(f_{n+2}) \\ &= r(f_1) + \beta P(f_1) V_\beta(T\pi) \end{aligned} \tag{2.7}$$

where $T\pi = (f_2, f_3, \ldots)$ is a strategy delayed one step for each time. Then, for any $N \times 1$ vector w, we define a function $L(f)$ that maps w into $L(f)w = r(f) + \beta P(f)w$. We may consider that a function $L(f)$ is a one-step return under the preceding return w when we use the decision $f \in F$. Thus, $V_\beta(f, \pi) = L(f)V_\beta(\pi)$.

We define *vector inequality* as follows. For any two column vectors w_1, w_2, we write $w_1 \geqslant w_2$ if every element of w_1 is not less than the corresponding element of w_2, and we write $w_1 > w_2$ if $w_1 \geqslant w_2$ and $w_1 \neq w_2$. This notation of vector inequality is used regularly.

DEFINITION 2.1.   *A strategy π^* is called β-optimal if $V_\beta(\pi^*) \geqslant V_\beta(\pi)$ for all π, where β ($0 \leqslant \beta < 1$) is fixed.*

This definition means that an optimal strategy is attained simultaneously for each initial state—a fact that is nontrivial, as will be shown later. If an optimal strategy π^* is attained simultaneously for each initial state, we have $aV_\beta(\pi^*) \geqslant aV_\beta(\pi)$ for any π and any initial distribution a; that is, an optimal strategy is independent of the initial distribution a. Conversely, if an optimal strategy that maximizes $aV_\beta(\pi)$ is independent of the initial distribution a, we have $V_\beta(\pi^*) \geqslant V_\beta(\pi)$.

LEMMA 2.2.   *$L(f)$ is monotone; that is, $w_1 \geqslant w_2$ implies $L(f)w_1 \geqslant L(f)w_2$.*

PROOF. $L(f)w_1 - L(f)w_2 = \beta P(f)(w_1 - w_2) \geqslant 0$ if $w_1 \geqslant w_2$. ∎

Under the above preparation, we have the following theorems:

THEOREM 2.3. $V_\beta(\pi^*) \geqslant V_\beta(g, \pi^*)$ for all $g \in F$ implies that π^* is β-optimal.

PROOF. The hypothesis is $L(g)V_\beta(\pi^*) \leqslant V_\beta(\pi^*)$ for all $g \in F$. For any strategy $\pi = (f_1, f_2, \ldots, f_n, \ldots)$, setting $g = f_n$, we have $L(f_n)V_\beta(\pi^*) \leqslant V_\beta(\pi^*)$. Applying the monotone operator $L(f_1) \cdots L(f_{n-1})$, we have

$$L(f_1)L(f_2) \cdots L(f_n)V_\beta(\pi^*) \leqslant L(f_1)L(f_2) \cdots L(f_{n-1})V_\beta(\pi^*).$$

Using this relation recursively, we have

$$V_\beta(\pi^*) \geqslant L(f_1)L(f_2) \cdots L(f_n)V_\beta(\pi^*) = V_\beta(f_1, \ldots, f_n, \pi^*) \qquad (2.8)$$

for all n. Letting $n \to \infty$, we have $V_\beta(\pi^*) \geqslant V_\beta(\pi)$ for all π. ∎

THEOREM 2.4. $V_\beta(f, \pi) > V_\beta(\pi)$ implies $V_\beta(f^\infty) > V_\beta(\pi)$.

PROOF. The hypothesis is $L(f)V_\beta(\pi) > V_\beta(\pi)$. Applying the monotone operator $L^{n-1}(f)$, we have $L^n(f)V_\beta(\pi) \geqslant L^{n-1}(f)V_\beta(f)$ for $n > 0$. Using this relation recursively, we have $L^n(f)V_\beta(\pi) \geqslant V_\beta(f, \pi) > V_\beta(\pi)$ for $n > 0$. Letting $n \to \infty$, we have $V_\beta(f^\infty) > V_\beta(\pi)$. ∎

Now we have our main theorem:

THEOREM 2.5. Take any $f \in F$. For each $i \in S$, denote $G(i, f)$ the set of all $k \in K_i$ for which

$$r_i^k + \beta \sum_{j \in S} p_{ij}^k v_j > v_i$$

where v_i is the ith element of $V_\beta(f^\infty)$. If $G(i, f)$ is empty for all $i \in S$, then f^∞ is β-optimal. For any $g \in F$ such that (a) $g(i) \in G(i, f)$ for some i and (b) $g(i) = f(i)$ whenever $g(i) \notin G(i, f)$, we have $V_\beta(g^\infty) > V_\beta(f^\infty)$.

PROOF. The ith element of $V_\beta(g, f^\infty)$ is $r_i^k + \beta \sum_{j \in S} p_{ij}^k v_j$, where $k = f(i)$. This will exceed v_i if and only if $g(i) \in G(i, f)$, and will equal v_i if $g(i) = f(i)$. Thus if $G(i, f)$ is empty for all $i \in S$, we have $V_\beta(f^\infty) \geqslant V_\beta(g, f^\infty)$ for all $g \in F$, which implies that f^∞ is β-optimal (by Theorem 2.3). While, for any $g \in F$ satisfying (a) and (b), we have $V_\beta(g, f^\infty) > V_\beta(f^\infty)$, so that $V_\beta(g^\infty) > V_\beta(f^\infty)$ (by Theorem 2.4). ∎

COROLLARY 2.6. There is a β-optimal strategy that is stationary.

PROOF. According to Theorem 2.5, if we take any stationary strategy f^∞, it is either β-optimal [case $G(i, f)$ empty for all $i \in S$], or it has a stationary improvement g^∞ [case $G(i, f)$ nonempty for some $i \in S$]. Since there are only finitely many stationary strategies, there is one that has no stationary improvement with finite iterations; hence it must be β-optimal. ∎

These theorems describe a method for finding an optimal stationary strategy, called the (Howard's) *policy iteration algorithm*. The algorithm has two parts as follows:

Value Determination Operation

Take any $f \in F$. Solve

$$v_i = r_i^k + \beta \sum_{j \in S} p_{ij}^k v_j$$

for $v_i (i \in S)$, where $k = f(i)$ corresponds to the chosen strategy f^∞.

Policy Improvement Routine

Using the values $v_i (i \in S)$, find the element of $G(i, f)$ for each $i \in S$ such that

$$r_i^k + \beta \sum_{j \in S} p_{ij}^k v_j > v_i$$

for all $k \in K_i$. If $G(i, f)$ is empty for all $i \in S$, f^∞ is β-optimal, and $V_\beta(f^\infty) = [v_i]$ is the discounted total expected return. If at least $g(i) \in G(i, f)$ for some i, we make an improved strategy g^∞ such that $g(i) \in G(i, f)$ for some i and $g(i) = f(i)$ for $G(i, f)$ empty; we then return to the value determination operation.

As an initial strategy f^∞, we may take, for example, $\max_{k \in K_i} r_i^k$ for each $i \in S$.

This policy iteration algorithm is simple and elegant. The numerical examples using this algorithm will be presented in Section 2.6. The convergence to an optimal strategy is fairly rapid. In Section 2.4 we shall discuss the setup of this algorithm from the viewpoint of linear programming.

2.3. Linear Programming Algorithm

In this section we consider the linear programming formulation of the discounted Markovian decision processes. (The relation between the policy iteration algorithm and a linear programming algorithm will be described in the next section.) We also formulate a linear programming algorithm, as well as discuss some interesting properties of this problem.

It is more convenient for subsequent discussion to extend the range of decisions to include *randomized* (or *mixed*) *strategies*. Thus for any n, we define $\lambda_i^k(n)$, the joint probability of being in state $i \in S$ and making decision $k \in K_i$. Our problem then is to find an optimal strategy that maximizes the discounted total expected return. Here we consider the maximization problem under the initial distribution (2.2) because an optimal strategy is attained simultaneously for each initial state.

Since $\lambda_i^k(n)$ obeys the probability law p_{ij}^k, we may write

$$\sum_{k \in K_j} \lambda_j^k(n) = \lambda_j(0) = a_j \quad \text{for} \quad n = 0$$

$$= \sum_{i \in S} \sum_{k \in K_i} p_{ij}^k \lambda_i^k(n-1) \quad \text{for} \quad n = 1, 2, \ldots, \quad j \in S \quad (2.9)$$

where a_j is the probability that the system is in state j at time 0, as defined in (2.2).

LEMMA 2.7. *Any nonnegative solution $\lambda_i^k(n)$ of (2.9) is a probability distribution and the corresponding discounted total expected return is bounded.*

PROOF. Since $\sum_{j \in S} p_{ij}^k = 1$, summing (2.9) over all $j \in S$ implies

$$\sum_{j \in S} \sum_{k \in K_j} \lambda_j^k(n) = \sum_{j \in S} a_j \quad \text{for} \quad n = 0$$

$$= \sum_{i \in S} \sum_{k \in K_i} \lambda_i^k(n-1) \quad \text{for} \quad n = 1, 2, \ldots \quad (2.10)$$

The assumption that a_j is a probability distribution implies recursively

$$\sum_{i \in S} \sum_{k \in K_i} \lambda_i^k(n) = 1 \quad \text{for} \quad n = 0, 1, 2, \ldots \quad (2.11)$$

This and the fact that $\lambda_i^k(n)$ is assumed to be nonnegative proves the first part of the lemma. The second part has been proved in (2.6). ∎

As a result of Lemma 2.7, we have the following objective function:

$$\max \sum_{n=0}^{\infty} \beta^n \sum_{i \in S} \sum_{k \in K_i} r_i^k \lambda_i^k(n) \quad (2.12)$$

under the constraints (2.9), and $\lambda_i^k(n) \geqslant 0$ for all $n = 0, 1, \ldots;$ $i \in S$, and $k \in K_i$. As it stands, this problem, say problem (P_0), is similar to a standard linear programming problem. However, it contains an infinite number of constraints and variables, and thus the classical theory of linear programming cannot be used to analyze it in this form. Using the fact that the sequences $\lambda_j^k(n)$ are bounded, and $0 \leqslant \beta < 1$, we define a set of new variables x_j^k by

$$x_j^k = \sum_{n=0}^{\infty} \beta^n \lambda_j^k(n) \quad \text{for} \quad j \in S, \quad k \in K_j \quad (2.13)$$

By definition, the new variables $x_j{}^k$ can be viewed as the z transforms of the sequences $\lambda_j{}^k(n)$ evaluated at $z = \beta$. Using these variables $x_j{}^k$, we have the following standard *linear programming problem*, say problem (P_T):

$$\max \sum_{j \in S} \sum_{k \in K_j} r_j{}^k x_j{}^k \tag{2.14}$$

subject to

$$\sum_{k \in K_j} x_j{}^k - \beta \sum_{i \in S} \sum_{k \in K_i} p_{ij}^k x_i{}^k = a_j \quad \text{for} \quad j \in S \tag{2.15}$$

$$x_j{}^k \geqslant 0 \quad \text{for} \quad j \in S, \quad k \in K_j \tag{2.16}$$

Problem (P_T) is now a standard linear programming problem. In the following discussion, we take advantage of the special structure of this problem to prove that its solution has several interesting properties. We shall then use these properties to show that there is always an optimal solution for this problem, and that for any basic optimal solution there is a corresponding optimal solution to problem (P_0).

We define a *stationary strategy* as a function that for each $i \in S$ selects exactly one variable $x_i{}^k$, where $k \in K_i$. This definition of stationary strategies will coincide with that of the preceding section. We shall now prove the following theorem for any stationary strategy.

THEOREM 2.8. *If the system of equations* (2.15) *is restricted to the variables* $x_i{}^k$ *selected by any stationary strategy, then* (a) *the corresponding subsystem has a unique solution;* (b) *if* $a_j \geqslant 0 \, (j \in S)$, *then* $x_j{}^k \geqslant 0 \, (j \in S)$; (c) *if* $a_j > 0 \, (j \in S)$ *then* $x_j{}^k > 0 \, (j \in S)$.

PROOF. The subsystem can be written

$$x_j - \beta \sum_{i \in S} p_{ij} x_i = a_j \quad \text{for} \quad j \in S \tag{2.17}$$

The superscript k is omitted because it is uniquely defined by the chosen stationary strategy.

Proof of (a). Consider the homogeneous system associated with (2.17) by assuming $a_j = 0 \, (j \in S)$. This system will certainly have at least a null solution. If x were another nonzero solution, define $I^- = \{i \mid x_i < 0\}$. Summing (2.5) over all $j \in I^-$ yields

$$\sum_{i \in I^-} [1 - \beta \sum_{j \in I^-} p_{ij}] x_i - \beta \sum_{i \notin I^-} \sum_{j \in I^-} p_{ij} x_i = \sum_{j \in I^-} a_j \tag{2.18}$$

The right-hand side of (2.18) is zero and all terms in the left-hand side are nonpositive, and thus must be zero. However, $\beta < 1$ implies $1 - \beta \sum_{j \in I^-} p_{ij} > 0$, which means that the first summation would be strictly negative. This is a contradiction except when $I^- = \phi$. By using an entirely similar argument, and defining $I^+ = \{i \mid x_i > 0\}$, we also conclude that $I^+ = \phi$. Thus the null solution is the unique solution of the homogeneous system. This proves that the rank of system (2.17) is N, and, therefore, that a unique solution exists for any value of the right-hand side a_j.

Proof of (b). Now let $a_j \geqslant 0$. The right-hand side of (2.18) is again nonnegative, and this implies $I^- = \phi$, that is, $x_j \geqslant 0$ ($j \in S$).

Proof of (c). Finally, let $a_j > 0$. In this case, we can define $I^- = \{i \mid x_i \leqslant 0\}$. The right-hand side of (2.18) would be strictly positive, while the left-hand side would be nonpositive. This implies that $I^- = \phi$ or $x_j > 0$. ∎

We shall now use Theorem 2.8 to prove an important relationship between stationary strategies and the basic feasible solutions of (2.15) when $a_j > 0$ ($j \in S$).

THEOREM 2.9.   *Whenever $a_j > 0$ ($j \in S$), there exists a one-to-one correspondence between stationary strategies and basic feasible solutions of (2.15). Moreover, any basic feasible solution is nondegenerate.*

PROOF.   Theorem 2.8 actually states that when $a_j > 0$ ($j \in S$), a stationary strategy has a corresponding unique solution of (2.17) that has N positive variables. This is, by definition, a basic nondegenerate feasible solution of (2.15). Conversely, if $x_i{}^k$ is a feasible solution of (2.15), we have

$$\sum_{k \in K_j} x_j{}^k = a_j + \beta \sum_{i \in S} \sum_{k \in K_i} p_{ij}^k x_i{}^k \geqslant a_j > 0 \qquad \text{for} \quad j \in S \qquad (2.19)$$

and thus, at least one variable $x_j{}^k$ has to be positive. Hence, there is exactly one $x_i{}^k > 0$ for each state j. This uniquely defines a stationary strategy. ∎

Any strategy associated in this way with an optimal basic solution will be called an *optimal stationary strategy*. In the next theorem we shall show that although an optimal stationary strategy is derived with a specific set of values $a_j > 0$, this strategy remains optimal for any nonnegative right-hand side a_j.

THEOREM 2.10.   *Whenever the right-hand side a_j of (2.15) is strictly positive, the problem (P_T) has an optimal basic solution and its dual has a unique optimal solution. Any optimal stationary strategy associated with it remains optimal for any nonnegative right-hand side of a_j.*

PROOF.   From Theorem 2.8, it is clear that there are feasible solutions, and we also know from Lemma 2.7 that the objective function is bounded. This guarantees the existence of an optimal solution for both the problem and its dual. By Theorem 2.9 any basic solution will be nondegenerate; and by the complementary slackness conditions, any optimal solution of the dual should satisfy the corresponding system of N dual equalities (which is nonsingular); therefore the dual solution is unique. The discussion of the dual problem will be stated in detail in the next section.

To prove the second part of the theorem, we note that the optimality of a given basic feasible solution of a linear programming problem depends on the objective function and not on the right-hand side. Changing the latter can only affect the feasibility for any nonnegative value of a_j. ■

The following corollary will be useful in the next section:

COROLLARY 2.11.   *For all positive right-hand side $a_j > 0$ (say $a_j = 1/N$), there exists a basic solution which has the property that for each $i \in S$, there is only one k such that $x_i^k > 0$ and $x_i^k = 0$ for k otherwise.*

PROOF.   This is a direct consequence of Theorems 2.9 and 2.10.

From the series of theorems, we know the special structure of the linear programming problem. Then with a suitable initial distribution (say, $a_j = 1/N$), we can get an optimal basic solution that corresponds to an optimal stationary strategy. We have presented two algorithms; one is the policy iteration algorithm discussed in the preceding section and the other is the linear programming algorithm discussed in this section. However, we cannot answer which is more advantageous in computing an optimal strategy. In the next section we shall discuss a relationship between the two algorithms.

2.4. Relationship between the Two Algorithms

In Section 2.2 we derived the policy iteration algorithm, which may be considered to be a successive approximation method in policy space of dynamic programming. In Section 2.3 we formulated the same problem by linear programming. In this section we show that these two algorithms are equivalent in mathematical programming.

We now require as the *primal problem* a linear programming problem discussed in the preceding section.

Primal problem:

$$\max \sum_{j \in S} \sum_{k \in K_j} r_j^k x_j^k \tag{2.20}$$

subject to

$$\sum_{k \in K_j} x_j^k - \beta \sum_{i \in S} \sum_{k \in K_i} p_{ij}^k x_i^k = a_j \qquad \text{for} \quad j \in S \tag{2.21}$$

$$x_j^k \geqslant 0 \qquad \text{for} \quad j \in S, \quad k \in K_j \tag{2.22}$$

The *dual problem* (2.20) is significant, for the reasons given in the preceding section.

Dual problem:

$$\min \sum_{i \in S} a_i v_i \tag{2.23}$$

subject to

$$v_i \geqslant r_i^k + \beta \sum_{j \in S} p_{ij}^k v_j \qquad \text{for} \quad i \in S, \quad k \in K_i, \tag{2.24}$$

$$v_i, \qquad \text{unconstrained in sign for} \quad i \in S \tag{2.25}$$

This dual problem can be immediately derived from the discussion of Section 2.2 as follows: For an optimal stationary strategy $\pi^* = f^\infty$, we have

$$V_\beta(f^\infty) \geqslant L(g)V_\beta(f^\infty) = r(g) + \beta P(g)V_\beta(f^\infty) \tag{2.26}$$

for all $g \in F$. Writing Eq. (2.26) element by element, we have the constraints (2.24). The objective function is the discounted total expected return starting in the initial distribution a:

$$aV_\beta(f^\infty) = \sum_{i \in S} a_i v_i \tag{2.27}$$

which implies the dual problem. It is more comprehensive to write the Tucker diagram in Table 2.1 for the primal and dual problems.

We now consider the solution of the primal problem. Since the constraints of the problem are equalities, we must use the two phase method or composite algorithms to obtain an initial basic feasible solution. But Corollary 2.11 implies that, for each $j \in S$, there is one k such that $x_j^k > 0$, and $x_j^k = 0$ otherwise. Thus, omitting phase I, we can obtain a basic feasible solution. To obtain an initial basic feasible solution, we use the usual simplex criterion for each $j \in S$. For example, we may apply

$$-r_j^* = \min_{k \in K_j} [-r_j^k] \qquad \text{for} \quad j \in S \tag{2.28}$$

or

$$r_j^* = \max_{k \in K_j} r_j^k \qquad \text{for} \quad j \in S \tag{2.29}$$

which corresponds to the initial strategy discussed in Section 2.2. Note that the asterisk stands for the data of a basic solution throughout this section.

TABLE 2.1

The Tucker Diagram for the Markovian Decision Process with Discounting

		Primal										Relations	Variables
	Variables	$x_1^1 > 0$	$x_1^2 > 0$	...	$x_2^1 > 0$	$x_2^2 > 0$	...	$x_N^1 > 0$	$x_N^2 > 0$	...			
Dual	v_1	$1 - \beta p_{11}^1$	$1 - \beta p_{11}^2$	...	$-\beta p_{21}^1$	$-\beta p_{21}^2$	...	$-\beta p_{N1}^1$	$-\beta p_{N1}^2$	...	$=$	a_1
	v_2	$-\beta p_{12}^1$	$-\beta p_{12}^2$	...	$1 - \beta p_{22}^1$	$1 - \beta p_{22}^2$	...	$-\beta p_{N2}^1$	$-\beta p_{N2}^2$	...	$=$	a_2
	.	.	.		.	.		.	.		.	.
	.	.	.		.	.		.	.		.	.
	v_{N-1}	$-\beta p_{1,N-1}^1$	$-\beta p_{1,N-1}^2$	...	$-\beta p_{2,N-1}^1$	$-\beta p_{2,N-1}^2$	...	$-\beta p_{N,N-1}^1$	$-\beta p_{N,N-1}^2$	...	$=$	a_{N-1}
	v_N	$-\beta p_{1N}^1$	$-\beta p_{1N}^2$	...	$-\beta p_{2N}^1$	$-\beta p_{2N}^2$	...	$1 - \beta p_{NN}^1$	$1 - \beta p_{NN}^2$	...	$=$	a_N
	Relations	$\wedge$	$\wedge$	...	$\wedge$	$\wedge$	...	$\wedge$	$\wedge$	...		
	Constants	r_1^1	r_1^2	...	r_2^1	r_2^2	...	r_N^1	r_N^2	...		

For any basic feasible solution, we have the basic matrix

$$
B = \begin{bmatrix}
1 - \beta p_{11}^* & -\beta p_{21}^* & \cdots & -\beta p_{N1}^* \\
-\beta p_{12}^* & 1 - \beta p_{22}^* & \cdots & -\beta p_{N2}^* \\
\cdot & \cdot & & \cdot \\
\cdot & \cdot & & \cdot \\
\cdot & \cdot & & \cdot \\
-\beta p_{1N}^* & -\beta p_{2N}^* & \cdots & 1 - \beta p_{NN}^*
\end{bmatrix} \tag{2.30}
$$

where p_{ji}^* is the transition probability according to the basic solution. Adding the row vector corresponding to the coefficients of the objective function, we have the modified basic matrix

$$
\bar{B} = \begin{bmatrix}
1 & -r_1^* & -r_2^* & \cdots & -r_N^* \\
0 & 1 - \beta p_{11}^* & -\beta p_{21}^* & \cdots & -\beta p_{N1}^* \\
0 & -\beta p_{12}^* & 1 - \beta p_{22}^* & \cdots & -\beta p_{N2}^* \\
\cdot & \cdot & \cdot & & \cdot \\
\cdot & \cdot & \cdot & & \cdot \\
\cdot & \cdot & \cdot & & \cdot \\
0 & -\beta p_{1N}^* & -\beta p_{2N}^* & \cdots & 1 - \beta p_{NN}^*
\end{bmatrix} = \begin{bmatrix} 1 & -r^{*T} \\ 0 & B \end{bmatrix} \tag{2.31}
$$

where we add the $(N + 1) \times 1$ unit vector with the first element unity to the first column. From the preceding section, B is nonsingular (Theorem 2.8); then so is $\bar{B}$. Thus, let $\bar{B}^{-1}$ be the inverse matrix of $\bar{B}$. We have

$$
\bar{B}^{-1} = \begin{bmatrix} 1 & \mu \\ 0 & B^{-1} \end{bmatrix} \tag{2.32}
$$

It is clear that $\bar{B}\bar{B}^{-1} = I$ (identity matrix). Hence, we have

$$
\mu = r^{*T} B^{-1} \tag{2.33}
$$

or

$$
B^T \mu^T = r^* \tag{2.34}
$$

where the superscript T denotes the transpose of the matrix. From its definition (2.32), μ is the vector of the *simplex multipliers* for the primal problem and is also the vector of the *dual variables*. Thus, we have

$$
\mu = (v_1, v_2, \ldots, v_N).
$$

Rewriting (2.34) element by element, we have

$$
v_i = r_i^* + \beta \sum_{j \in S} p_{ij}^* v_j \qquad \text{for} \quad j \in S \tag{2.35}
$$

which corresponds to the *value determination operation* in the policy iteration algorithm discussed in Section 2.2. In other words, solving the system of N linear equations (2.35) is equivalent to finding the simplex multipliers (and also the dual variables) for the primal problem. (This derivation is straight-forward from the *complementary slackness principle*.)

Next, consider the *simplex criterion* for the next step using these simplex multipliers. Let the simplex multiplier for the objective function be 1, and write

$$\bar{\mu} = (1, v_1, v_2, \ldots, v_N)$$

The coefficients of the corresponding linear programming problem can be written by

$$
\bar{A} =
\begin{bmatrix}
-r_1{}^1 & -r_1{}^2 & \cdots & -r_N{}^1 & -r_N^{K_N} \\
-\beta p_{11}^1 & 1 - \beta p_{11}^2 & \cdots & -\beta p_{N1}^1 & -\beta p_{N1}^{K_N} \\
-\beta p_{12}^1 & -\beta p_{12}^2 & \cdots & -\beta p_{N2}^1 & -\beta p_{N2}^{K_N} \\
\cdot & \cdot & & \cdot & \cdot \\
\cdot & \cdot & & \cdot & \cdot \\
\cdot & \cdot & & \cdot & \cdot \\
-\beta p_{1N}^1 & -\beta p_{1N}^2 & \cdots & 1 - \beta p_{NN}^1 & 1 - \beta p_{NN}^{K_N}
\end{bmatrix}
\tag{2.36}
$$

The simplex criterion of the next step is

$$
\begin{aligned}
[\Delta_i{}^k] &= \bar{\mu}\bar{A} \\
&= \left[-r_i^k - \beta \sum_{j \in S} p_{ij}^k v_j + v_i \right]
\end{aligned}
\tag{2.37}
$$

It is evident that, for the basic variables,

$$\Delta_i{}^* = -r_i{}^* - \beta \sum_{j \in S} p_{ij}^* v_j + v_i = 0 \qquad \text{for} \quad i \in S \tag{2.38}$$

If for all $i \in S$ and $k \in K_i$,

$$\Delta_i{}^k = -r_i{}^k - \beta \sum_{j \in S} p_{ij}^k v_j + v_i \geqslant 0,$$

or by using (2.38), for all $i \in S$ and $k \in K_i$,

$$r_i{}^* + \beta \sum_{j \in S} p_{ij}^* v_j \geqslant r_i{}^k + \beta \sum_{j \in S} p_{ij}^k v_j \tag{2.39}$$

we have an optimal solution from the basic theory of linear programming.

If there is at least one pair $i \in S$ and $k \in K_i$ such that

$$\Delta_i{}^k = -r_i{}^k - \beta \sum_{j \in S} p_{ij}^k v_j + v_i < 0 \tag{2.40}$$

or by using (2.38),

$$r_i{}^* + \beta \sum_{j \in S} p_{ij}^* v_j < r_i{}^k + \beta \sum_{j \in S} p_{ij}^k v_j \tag{2.41}$$

there exists an improved solution, or an improved strategy, that corresponds to the *policy improvement routine* in the policy iteration algorithm.

Consequently, the policy iteration algorithm is only a special extension of linear programming which has the property that pivot operations for many (at most N) variables are performed simultaneously. We have already seen in Section 2.2 that the substitutions for many variables imply an improved strategy. But even if there is only one pair $i \in S$ and $k \in K_i$ satisfying (2.41), we must solve the system of N linear equations (2.35)—a computational disadvantage. Questions of this nature, which arise in computing an optimal strategy, will be discussed in Section 3.6.

2.5. Return Structures

In the preceding discussion we assumed that we received the return when the system was in state i.

We extend the return structures as follows: When the system is in state i, two things happen: (a) We receive the return r_i'; (b) if the system moves to state j at the next time, we receive the return r_{ij}'.

Under the above return structures, $v_i(n)$, the n-step total expected return, is

$$v_i(n) = r_i' + \sum_{j \in S} p_{ij}[r_{ij}' + \beta v_j(n - 1)]$$
$$= r_i' + \sum_{j \in S} p_{ij}r_{ij}' + \beta \sum_{j \in S} p_{ij}v_j(n - 1) \qquad (2.42)$$

Substituting

$$r_i = r_i' + \sum_{j \in S} p_{ij}r_{ij}' \qquad (2.43)$$

we have the same return structures treated in the preceding discussion. In what follows (up to Chapter 5), we consider only the return r_i^k for the discrete-time models.

2.6. Examples

This section gives two numerical examples and their solutions by using two algorithms, the policy iteration algorithm and a linear programming algorithm.

EXAMPLE 1.   The first example was given in Chapter 1. Now we define $F = \{f_1, f_2\}$, where $f_k(1) = 1$ and $f_k(2) = k$.

Then

$$P(f_1) = \begin{bmatrix} 0.7 & 0.3 \\ 0.6 & 0.4 \end{bmatrix}, \qquad r(f_1) = \begin{bmatrix} 3 \\ -2 \end{bmatrix}$$

$$P(f_2) = \begin{bmatrix} 0.7 & 0.3 \\ 0.4 & 0.6 \end{bmatrix}, \qquad r(f_2) = \begin{bmatrix} 3 \\ -1 \end{bmatrix}$$

where f_1 denotes a strategy of a rapid repair and f_2 a strategy of a usual repair. We now solve this problem setting a discount factor $\beta = 0.9$.

First, we apply the policy iteration algorithm. Take an initial strategy f_1. From the value determination operation, we have

$$v_1 = \tfrac{1380}{91}, \qquad v_2 = \tfrac{880}{91}$$

By using the values obtained above and the policy improvement routine, we obtain

$$r_2{}^2 + \beta \sum_{j=1}^{2} p_{2j}^2 v_j = \tfrac{881}{91} > \tfrac{880}{91} = v_2$$

which implies an improved strategy f_2. Returning to the value determination operation, we have

$$v_1 = \tfrac{1110}{73}, \qquad v_2 = \tfrac{710}{73}$$

which concludes with a 0.9-optimal f_2.

Second, we apply the linear programming approach. Under an initial distribution

$$a = (0.5, 0.5)$$

we have the following linear programming problem:

$$\max [3x_1{}^1 - 2x_2{}^1 - x_2{}^2]$$

subject to

$$0.37x_1{}^1 - 0.54x_2{}^1 - 0.36x_2{}^2 = 0.5$$
$$-0.27x_1{}^1 + 0.64x_2{}^1 + 0.46x_2{}^2 = 0.5$$
$$x_1{}^1, x_2{}^1, x_2{}^2 \geqslant 0$$

Thus, the optimal solution is

$$x_1{}^1 = \tfrac{410}{73}, \qquad x_2{}^2 = \tfrac{320}{73}$$

and the objective function is $\tfrac{910}{73}$. The value of our objective function coincides with

$$aV_\beta(f_2{}^\infty) = \begin{bmatrix} 0.5 & 0.5 \end{bmatrix} \begin{bmatrix} \tfrac{1110}{73} \\ \tfrac{710}{73} \end{bmatrix} = \tfrac{910}{73}$$

EXAMPLE 2.   (Howard's taxicab problem [63, p. 44].) Consider the problem of a taxicab driver whose territory encompasses three towns, A, B and C. If he is in town A, he has three possible courses of action: (1) He can cruise in the hope of picking up a passenger by being hailed. (2) He can drive to the nearest cab stand and wait in line. (3) He can pull over and wait for a radio car.

If he is in town C, he has the same three alternatives, but if he is in town B, the last action is not present because there is no radio-car service in that town. For a given town and a given action, there is a probability that the next trip will be to each of the towns A, B, and C and that there will be a corresponding return in monetary units with each trip. This return presents the income from the trip after all necessary expenses have been deducted. For example, in the case of actions 1 and 2, the cost of cruising or of driving to the nearest cab stand must be included in calculating the returns. The probabilities of transition and the returns depend upon the action, since different customers will be encountered under each action.

If we identify towns A, B, and C with states 1, 2, and 3, respectively, then Table 2.2 results.

TABLE 2.2

Data for the Taxicab Problem

State i	Action k	Probability p_{i1}^k	p_{i2}^k	p_{i3}^k	Return r_{i1}^k	r_{i2}^k	r_{i3}^k	$r_i^k = \sum_{j=1}^{3} p_{ij}^k r_{ij}^k$
	1	$\frac{1}{2}$	$\frac{1}{4}$	$\frac{1}{4}$	10	4	8	8
1	2	$\frac{1}{16}$	$\frac{3}{4}$	$\frac{3}{16}$	8	2	4	2.75
	3	$\frac{1}{4}$	$\frac{1}{8}$	$\frac{5}{8}$	4	6	4	4.25
2	1	$\frac{1}{2}$	0	$\frac{1}{2}$	14	0	18	16
	2	$\frac{1}{16}$	$\frac{7}{8}$	$\frac{1}{16}$	8	16	8	15
	1	$\frac{1}{4}$	$\frac{1}{4}$	$\frac{1}{2}$	10	2	8	7
3	2	$\frac{1}{8}$	$\frac{3}{4}$	$\frac{1}{8}$	6	4	2	4
	3	$\frac{3}{4}$	$\frac{1}{16}$	$\frac{3}{16}$	4	0	8	4.5

From R. A. Howard, "Dynamic Programming and Markov Processes." M.I.T. Press, Cambridge, Massachusetts, 1960.

Here we assume $\beta = 0.90$. Let an initial policy be

$$f = \begin{bmatrix} 1 \\ 1 \\ 1 \end{bmatrix}$$

which is derived from $\max_{k \in K_i} r_i^k$ for each $i \in S$.

The policy iteration algorithm yields an optimal strategy. The calculations of the policy iteration algorithm are given in Table 2.3.

TABLE 2.3

Solution of the Taxicab Problem by the Policy Iteration Algorithm, with $\beta = 0.9$

	VDO^a	PIV^b	VDO	PIV	VDO	PIV	
v_1	91.3		119.4		121.7		
v_2	97.6		134.5		135.3		
v_3	90.0		121.9		122.8		
f	1	1		2		2	
	1	2		2		2	STOP
	1	2		2		2	

^{a} VDO is the value determination operation.
^{b} PIV is the policy improvement routine.

We can also solve this problem by linear programming. We omit the solution.

2.7. Sensitivity Analysis with Respect to a Discount Factor

In the preceding discussion we considered the discounted problem for a fixed discount factor β. In this section we study the *sensitivity analysis* with respect to a discount factor, that is, how the optimal strategy changes when β varies between 0 and 1. This analysis has a close relation to 1-optimal strategies in the nondiscounted Markovian decision model, which will be given in Chapter 4.

EXAMPLE. (Veinott [111].) There are two states, 1 and 2. In state 1 there are three actions $k = 1, 2, 3$. Action k yields the return of $(5 - k)/4$ and the system remains in state 1 with probability $(k - 1)/2$. In state 2 there is only one action that yields 0, and the system remains in state 2 with certainty. Now $F = \{f_1, f_2, f_3\}$, where $f_k(1) = k$ and $f_k(2) = 1$ $(k = 1, 2, 3)$. Then

$$P(f_k) = \begin{bmatrix} (k-1)/2 & (3-k)/2 \\ 0 & 0 \end{bmatrix}, \qquad r(f_k) = \begin{bmatrix} (5-k)/4 \\ 0 \end{bmatrix}$$

Thus we have

$$V_\beta(f_k) = \begin{bmatrix} \dfrac{(5-k)}{1 - (k-1)\beta/2} \\ 0 \end{bmatrix}$$

for $k = 1, 2, 3$. Then we can write $v_1(f_k)$ $(k = 1, 2, 3)$ for $\beta \in [0, 1]$ in Fig. 2.1.

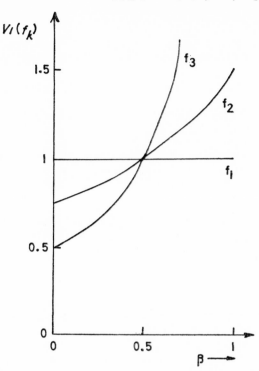

FIG. 2.1. The discounted total expected return $v_1(f_k)$ as a function of discount factor $\beta \in [0, 1]$.

For this problem, f_1 is β-optimal for $\beta \in [0, \frac{1}{2}]$, f_2 is $\frac{1}{2}$-optimal and f_0 is β-optimal for $\beta \in [\frac{1}{2}, 1]$. Note that f_1, f_2, and f_3 are all $\frac{1}{2}$-optimal. The optimal strategy regions as a function of the discount factor β for Veinott's example is shown below:

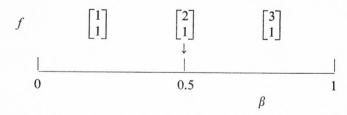

Second, we consider Example 2 of the preceding section (taxicab problem).

Howard [63, p. 88] has given the optimal strategy regions as a function of the discount factor for the taxicab problem:

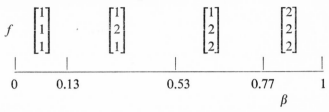

We note from the above diagrams that there exists a point β_0 such that two or three strategies are all optimal. This point is called a *critical point* [63] or an *indifference point* [106]. For instance, $\beta_0 = \frac{1}{2}$ is an indifference point for the first example. We also note that there are two strategies that differ in only a single state for which β_0 is an indifference point. We can consider, however, a case when there are two strategies that differ in two or more states for which β_0 is an indifference point. The following is such an example, but it is impossible. That is, we have the following:

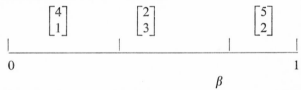

THEOREM 2.12. *If β_0 is an indifference point, then there must be two stationary strategies that differ in only a single state for which β_0 is an indifference point.*

PROOF. Assume that two optimal strategies f and f' have an indifference point at $\beta = \beta_0$. If P, r and P', r' are the respective Markov matrices and the return vectors for the two strategies, then V [abbreviated $V_\beta(f)$] must be the same for the two strategies at their indifference point β_0. Hence we have

$$V = r + \beta_0 P V = r' + \beta_0 P' V \tag{2.44}$$

or

$$v_i = r_i + \beta_0 \sum_{j \in S} p_{ij} v_j = r_i' + \beta_0 \sum_{j \in S} p_{ij}' v_j \quad \text{for} \quad i \in S \tag{2.45}$$

where $V = [v_i]$, $V' = [v_i']$, and so on. Now if we construct a new strategy, each of whose elements comes from one of the corresponding elements in f of f', the corresponding P'' and r'' will also satisfy (2.44). Hence this new strategy f'' shares the indifference point β_0 with f and f'. Thus if f and f' differ by more than a single element (for example, $f = [\frac{1}{3}]$, $f' = [\frac{2}{5}]$), then there are other strategies that share this indifference point with f and f' (for example, both $f'' = [\frac{1}{5}]$ and $f''' = [\frac{2}{3}]$) and only differ with f or f' in a single element. ∎

Here we omit an algorithm for finding indifference points. For a detailed discussion of this algorithm, see Smallwood [106].

We have an interesting problem of the behaviour $V_\beta(f^\infty)$ as $\beta \to 1$. This problem will be mentioned in detail in Chapter 4.

References and Comments

Markov chain theory is found in many standard texts on probability theory or Markov chain theory. See, for example, Kemeny and Snell [70], Feller [56], Bharucha-Reid [12] Karlin [68], Doob [50], and Chung [23].

A Markovian decision process with discounting was first discussed by Howard [63]. The policy iteration algorithm was also introduced by Howard [63]. Blackwell [14] has studied rigorously the same process and has shown that there exists a β-optimal stationary strategy (Corollary 2.6). Blackwell [14] has also given Lemma 2.2 and Theorems 2.3, 2.4, and 2.5.

Linear programming formulation of the discounted process was first given by D'Epenoux [48]. The discussion in Section 2.3 was given by De Ghellinck and Eppen [28]. The general theory of linear programming is found in Dantzig [25], for example.

Example 2 in Section 2.6 has been given by Howard [63], who has also given other interesting examples, for example, the baseball problem and the automobile replacement problem. In Section 3.6 we shall give the automobile replacement problem as an example of the nondiscounted models.

Theorem 2.12 has been given by Smallwood [106], who has also given an algorithm for finding indifference points by using the so-called Cayley–Hamilton theorem in linear algebra. We omitted this result from our discussion.

We now describe some interesting problems that we previously omitted. The first problem is that of selecting an initial strategy. The rapidity of convergence to a β-optimal strategy depends on the selection of an initial strategy, regardless of the algorithms we apply. We proposed $\max_{k \in K_i} r_i^k$ for each $i \in S$ as an initial strategy from the viewpoint of linear programming.

MacQueen [80] has studied, from the viewpoint of dynamic programming, another successive approximation approach for converging monotonely to a β-optimal strategy and has also shown [81] the efficiency of his algorithm.

The second problem is a special Markovian decision process such that

$$p_{ij}^k = p_{\cdot j}^k, \qquad r_i^k = W_i + V^k$$

and the process satisfies the nested condition. This problem is called a separable Markovian decision process [28]. Howard's automobile replacement problem (which will be shown in Section 3.6) is a separable Markovian decision problem. For the process, De Ghellinck and Eppen [28], and Denardo [32] have studied and given the simple algorithms.

The third problem is the so-called turnpike theorem, which states that a β-optimal strategy when the horizon is sufficiently large is to choose one of strategies which is β-optimal when the horizon is infinite. Shapiro [104] has proven this turnpike theorem and has also given an upper bound of the horizon.

Chapter 3

Markovian Decision Processes
with No Discounting   I

3.1. Introduction

We consider Markovian decision processes with no discounting in this and the next chapters. We only treat processes over an infinite planning horizon. In this process, the total expected return will be usually divergent. (In Section 3.8 we consider a special case in which the total expected return is finite.) Thus, there will not be the elegant discussion of the preceding chapter, since the Markov chain under consideration may change its state classification structure from strategy to strategy.

We shall separate the discussion of Markovian decision processes with no discounting into two chapters. In this chapter we consider special structure processes, that is, the *completely ergodic process* and the *terminating process*. In the next chapter we shall consider the general Markovian decision process in which the state classification may change from strategy to strategy.

Two approaches have been considered for Markovian decision processes with no discounting. One is treating $\beta = 1$ as a limiting case of $\beta < 1$, where we can use the discussion of the preceding chapter. Another is that of treating directly the long-run average return per unit time. We mainly use the first approach, with a brief treatment of the second.

3.2. Preliminaries

This section provides the preliminaries for this and the following chapters. The results of this section will supply the fundamental properties for later discussion.

First, we cite some well-known facts concerning matrices.

LEMMA 3.1.   *For any $N \times N$ matrix A, if A^n tends to 0 (zero matrix) as n tends to infinity, then $(I - A)$ is nonsingular, and*

$$(I - A)^{-1} = \sum_{i=0}^{\infty} A^i \qquad (3.1)$$

The proof of this lemma will be found elsewhere (see, for example, Kemeny and Snell [70, p. 22]).

LEMMA 3.2.   *For any $N \times N$ Markov matrix P, if $\sum_{i=0}^{n-1} P^i/n$ tends to P^* as n tends to infinity, then $(1 - \beta) \sum_{i=0}^{\infty} \beta^i P^i$ tends to P^* as $\beta \to 1 - 0$.*

PROOF.   The hypothesis asserts Césaro summability of $\{P^n\}$ to P^*. Thus Césaro summability of $\{P^*\}$ to P^* implies Abel summability of $\{P^n\}$ to P^*, that is,

$$(1 - \beta) \sum_{i=0}^{\infty} \beta^i P^i \to P^* \qquad \text{as} \quad \beta \to 1 - 0. \blacksquare \tag{3.2}$$

The next lemma will be important for later discussion:

LEMMA 3.3.   *Let P be any $N \times N$ Markov matrix.* (a) *The sequence $\sum_{i=0}^{n-1} P^i/n$ converges as $n \to \infty$ to a Markov matrix P^* such that*
$$PP^* = P^*P = P^*P^* = P^* \tag{3.3}$$
(b) $I - (P - P^*)$ *is nonsingular, and*
$$H(\beta) = \sum_{i=0}^{\infty} \beta^i(P^i - P^*) \to H = (I - P + P^*)^{-1} - P^* \tag{3.4}$$
as $\beta \to 1 - 0$.
$$H(\beta)P^* = P^*H(\beta) = HP^* = P^*H = 0 \tag{3.5}$$
and
$$(I - P)H = H(I - P) = I - P^* \tag{3.6}$$
(c) rank $(I - P)$ + rank $P^* = N$ \tag{3.7}
(d) *For every $N \times 1$ column vector c, the system*
$$Px = x, \qquad P^*x = P^*c \tag{3.8}$$
has a unique solution x.

PROOF.   (a) may be found in a text on Markov chains (for example, Kemeny and Snell [70]).

Proof of (b).   Using (3.3), we have $P^i - P^* = (P - P^*)^i$ for $i > 0$. From Lemma 3.1,

$$H(\beta) = \sum_{i=0}^{\infty} \beta^i(P - P^*)^i - P^* = [I - \beta(P - P^*)]^{-1} - P^*$$

for $0 \leqslant \beta < 1$, or $(H(\beta) + P^*)(I - \beta(P - P^*)) = I$. That is,
$$(H(\beta) + P^*)(I - P + P^*) = I - (1 - \beta)H(\beta)(P - P^*) \tag{3.9}$$
From Lemma 3.2, we have

$$(1 - \beta) \sum_{i=0}^{\infty} \beta^i(P^i - P^*) = (1 - \beta)H(\beta) \to 0 \tag{3.10}$$

as $\beta \to 1 - 0$. Thus the matrix on the right-hand side of (3.9) tends to I as $\beta \to 1 - 0$, and $(I - P + P^*)$ is nonsingular. Postmultiplying (3.9) by $(I - P + P^*)^{-1}$ and letting $\beta \to 1 - 0$ yields $H(\beta) + P^* \to (I - P + P^*)^{-1}$, which proves (3.4). For (3.5), we have

$$\begin{aligned} H(\beta)P^* &= [(I - \beta(P - P^*))^{-1} - P^*]P^* \\ &= [I - \beta(P - P^*)]^{-1}[I - (I - \beta(P - P^*))P^*]P^* \\ &= [I - \beta(P - P^*)]^{-1}[P^* - P^* + \beta P^* - \beta P^*] = 0 \end{aligned}$$

from (3.3). We can show by a similar proof that $P^*H(\beta) = HP^* = P^*H = 0$. For (3.6), we have

$$\begin{aligned} (I - P)H &= [(I - P + P^*) - P^*]H = (I - P + P^*)H \\ &= (I - P + P^*)[(I - P + P^*)^{-1} - P^*] = I - P^* \end{aligned}$$

from (3.3) and (3.5). We can also show that $H(I - P) = 0$.

Proof of (c). It follows that

$$N = \text{rank } (I - P + P^*) \leqslant \text{rank } (I - P) + \text{rank } P^* \leqslant N.$$

Proof of (d). The proof is a straightforward consequence of (c), thus completing the proof of the lemma. ∎

Following the notation of the preceding chapter, the total expected return up to time n starting in each state is

$$\sum_{i=0}^{n-1} P_i(\pi)r(f_{i+1}) \tag{3.11}$$

for any strategy $\pi = (f_1, f_2, \ldots)$. Thus the limit *infimum* of the average return per unit time as n tends to infinity is

$$\Gamma(\pi) = \lim_{n \to \infty} \inf \frac{1}{n} \sum_{i=0}^{n-1} P_i(\pi)r(f_{i+1}) \tag{3.12}$$

Our problem is then to find a strategy that maximizes (3.12) under all strategies π—that is, an *optimal strategy* π^* such that

$$\Gamma(\pi^*) \geqslant \Gamma(\pi) \tag{3.13}$$

for all π.

Derman [36] has shown that there is an optimal strategy that is stationary, where an optimal strategy is considered under the average return criterion. The same result has also been proved by Blackwell [14] and others. Here we demonstrate Derman's result [36].

THEOREM 3.4. *There is an optimal strategy that is stationary.*

PROOF. We have seen that there is a β-optimal stationary strategy π^* for the discounted Markovian decision process (Corollary 2.6). Since the number of stationary strategies is finite, it is possible to choose a sequence $\{\beta_v\}$, $\lim_{v \to \infty} \beta_v = 1$, such that $\pi^* = \pi_{\beta_v}$ ($v = 1, 2, \ldots$), where $\pi_{\beta_v} = f^{*\infty}$ is a stationary strategy. From Lemma 3.2, we have

$$\lim_{v \to \infty} (1 - \beta_v) \sum_{i=0}^{\infty} \beta_v{}^i P^i(f_{\beta v}) r(f_{\beta v}) = P^*(f^*) r(f^*) = u(f^*) \tag{3.14}$$

For any strategy π we have (using Widder [119, p. 181]):

$$\liminf_{n \to \infty} \frac{1}{n} \sum_{i=0}^{n-1} P_i(\pi) r(f_{i+1}) \leqslant \lim_{v \to \infty} (1 - \beta_v) V_{\beta v}(\pi)$$

$$\leqslant \lim_{v \to \infty} (1 - \beta_v) V_{\beta v}(\pi_{\beta v}) = u(f^*) \tag{3.15}$$

where $V_\beta(\pi)$ is the discounted total expected return for a discount factor β ($0 \leqslant \beta < 1$). Thus, there is an optimal strategy $f^{*\infty}$ that is stationary. ∎

This theorem shows that there is an optimal stationary strategy under the average criterion, where an optimal strategy is attained simultaneously for all initial states. Thus, an optimal strategy remains optimal for any initial distribution α.

In the sequel (up to Section 3.6), we shall discuss a problem of special structure, the so-called completely ergodic process.

DEFINITION 3.5. *For a Markovian decision process, the Markov chain under consideration is always ergodic regardless of the strategies we choose. The process is called the* completely ergodic process.

The examples presented in Section 2.6 are all completely ergodic processes. There are, however, many examples that are not completely ergodic. We discuss only the completely ergodic processes throughout this chapter (except in Section 2.7). The general case where there are several ergodic sets and some transient states will be considered in the next chapter.

Note that the process, which extends the range of decisions to include randomized strategies, is also completely ergodic from its definition.

For any stationary strategy f^∞ of the completely ergodic process, we have

$$\liminf_{n \to \infty} \frac{1}{n} \sum_{i=0}^{n-1} P_i(\pi) r(f_{i+1}) = \lim \frac{1}{n} \sum_{i=0}^{n-1} P^i(f) r(f) = P^*(f) r(f) \tag{3.16}$$

where P^* is the limiting matrix that is composed of all identical rows $\pi(f) = [\pi_j(f)]$. That is,

$$P^*(f) = 1\pi(f) \tag{3.17}$$

where $\mathbf{1}$ is the $N \times 1$ column vector with all elements 1. Then, the $N \times 1$ row vector $\pi(f)$ is a unique solution of

$$\pi(f) = \pi(f)P(f) \tag{3.18}$$

$$\pi_j(f) > 0 \qquad \text{for} \quad j \in S \tag{3.19}$$

$$\pi(f)\mathbf{1} = 1 \tag{3.20}$$

Thus an average return per unit time for a stationary strategy starting in each state is

$$P^*(f)r(f) = \mathbf{1}\pi(f)r(f) = (\sum_{j \in S} \pi_j(f)r_j^k)\mathbf{1} \tag{3.21}$$

where $k = f(j)$. This means that the average return per unit time is the weighted sum of r_j^k over the limiting probabilities $\pi_j(f)$ and is identical for any initial state $i \in S$. That is, the average return is independent of any initial distribution.

We have seen from the preceding discussion that our problem is to find an optimal strategy within the finite set of stationary strategies, since there is an optimal stationary strategy. Our problem is thus a combinatorial one, and we can apply the direct enumeration method. (This is valid not only for the completely ergodic process but also for the general process in Chapter 4.) In most cases, however, the direct enumeration approach seems to be impossible because the number of stationary strategies, $K_1 \times K_2 \times \cdots \times K_N$, is too numerous. Thus we shall need efficient algorithms to find an optimal strategy.

3.3. Policy Iteration Algorithm

In this section we shall establish a policy iteration algorithm for the completely ergodic process. The discussion of this section presents a special case of the results of Chapter 4; the assumption of complete ergodicity makes the discussion of this section simple.

For any stationary strategy f^∞, $V^n(f)$, the total expected return vector up to time n, satisfies the following recursive relation:

$$V^n(f) = r(f) + P(f)V^{n-1}(f) \qquad \text{for} \quad n = 1, 2, \ldots \tag{3.22}$$

where

$$V^0(f) = 0 \tag{3.23}$$

Throughout this and the following chapters, we shall restrict our attention to stationary strategies; we shall write as a stationary strategy f instead of f^∞.

LEMMA 3.6. *If $P^n(f)$ tends to $P^*(f)$ as n tends to infinity, that is, $P(f)$ is regular, we have*

$$V^n(f) = n g \mathbf{1} + v(f) + \varepsilon(n, f) \qquad \text{for} \quad n > 0 \tag{3.24}$$

where $\varepsilon(n, f)$ tends to zero as n tends to infinity, and

$$g = \pi(f)r(f), \quad and \quad v(f) = H(f)r(f).$$

PROOF.

$$V^n(f) = \sum_{i=0}^{n-1} P^i(f)r(f)$$

$$= nP^*(f)r(f) + [\sum_{i=0}^{n-1} (P(f) - P^*(f))^i - P^*(f)]r(f)$$

$$= nP^*(f)r(f) + [\sum_{i=0}^{\infty} (P(f) - P^*(f))^i - P^*(f)]r(f)$$

$$- \sum_{i=n}^{\infty} (P(f) - P^*(f))^i r(f) \qquad (3.25)$$

Setting

$$g1 = P^*(f)r(f) = \pi(f)r(f)1$$
$$v(f) = H(f)r(f)$$

and

$$\varepsilon(n, f) = - \sum_{i=n}^{\infty} (P(f) - P^*(f))^i r(f). \blacksquare$$

Note that in this lemma, if we set $u(f) = P^*(f)r(f)$, then (3.24) is valid for the general case, which will be discussed in Chapter 4.

For a sufficiently large n, substituting $V^n(f)$ for $ng1 + v(f)$ in (3.22), we have

$$ng1 + v(f) = r(f) + (n - 1)g1 + P(f)v(f) \qquad (3.26)$$

or

$$g1 + v(f) = r(f) + P(f)v(f) \qquad (3.27)$$

Then we note that $v(f)$ in (3.27) is a relative value, that is, $v(f) + c1$ (c, constant) also satisfies (3.27). Hence, setting one element of $v(f)$ as zero (for example, $v_N = 0$), we can solve (3.27) with respect to g and $v(f)$. This corresponds to the *value determination operation* for the completely ergodic process.

Next, we shall need the *policy improvement routine*. Here we shall apply the heuristic approach of Howard [63]. For a sufficiently large n, the right-hand side of (3.22) is asymptotically

$$r(f) + P(f)((n - 1)g1 + v(f)) \qquad (3.28)$$

Noting $\sum_{j \in S} p_{ij} = 1$, we have

$$r(f) + P(f)v(f) + (n - 1)g1 \qquad (3.29)$$

For the above equation, $(n - 1)g1$ is no contribution to the next improved strategy. Hence we may maximize the remaining terms

$$r(f) + P(f)v(f) \qquad (3.30)$$

with respect to all actions in each state using the known values $v(f)$. The rigorous treatment of the policy improvement routine will be discussed in Section 3.5.

If there is a strategy f' such that

$$r(f') + P(f')v(f) > r(f) + P(f)v(f) = v(f) \qquad (3.31)$$

we can obtain an improved strategy f', where the vector inequality of (3.31) is that of Section 2.2.

THEOREM 3.7.   *If* $r(f') + P(f')v(f) > r(f) + P(f)v(f)$ *for some* $f' \in F$, *we have* $g(f') > g(f)$, *where* $g(f)$ *denotes the average return per unit time using a stationary strategy* f.

PROOF.   Let γ be the $N \times 1$ column such that

$$\gamma = r(f') + P(f')v(f) - r(f) - P(f)v(f) > 0$$

For two strategies f', f, we have

$$g(f')\mathbf{1} + v(f') = r(f') + P(f')v(f') \qquad (3.32)$$

$$g(f)\mathbf{1} + v(f) = r(f) + P(f)v(f) \qquad (3.33)$$

Substracting (3.33) from (3.32), and setting

$$\Delta g = g(f') - g(f), \qquad \Delta v = v(f') - v(f),$$

we have

$$\Delta g\mathbf{1} + \Delta v = \gamma + P(f)\Delta v \qquad (3.34)$$

Premultiplying the limiting vector $\pi(f)$ of $P(f)$, we have

$$\Delta g \quad [= \pi(f)\Delta g\mathbf{1}] = \pi(f)\gamma > 0 \qquad (3.35)$$

since $\pi_j(f)$ is positive for all $j \in S$ from the assumption of complete ergodicity. Thus we can obtain an improved strategy f'. ∎

We summarize the *policy iteration algorithm* for the completely ergodic process as follows:

Value Determination Operation

Take any stationary f^∞. Solve

$$g + v_i = r_i^k + \sum_{j=1}^{N-1} p_{ij}^k v_j$$

for $g, v_1, v_2, \ldots, v_{N-1}$ (setting $v_N = 0$), where the superscript k is determined by the chosen strategy f^∞.

Policy Improvement Routine

Using the values v_i and g, find the element of $G(i, f)$ for each $i \in S$ such that

$$r_i^k + \sum_{j=1}^{N-1} p_{ij}^k v_j > \mathfrak{g} + v_i$$

for all $k \in K_i$. If $G(i, f)$ is empty for all $i \in S$, f^∞ is optimal, $\mathfrak{g}$ is the average return per unit time, and $v_1, \ldots, v_{N-1}$ are the relative bias terms. If at least $g(i) \in G(i, f)$ for some i, make an improved strategy g^∞ such that $g(i) \in G(i, f)$ for some i and $g(i) = f(i)$ for $G(i, f)$ empty; then return to the value determination operation.

In the policy improvement routine, if there are two or more actions satisfying $G(i, f)$ for some $i \in S$, we should apply an improved strategy $g(i)$ such that

$$\max_{k \in K_i} \left[r_i^k + \sum_{j=1}^{N-1} p_{ij}^k v_j \right] \tag{3.36}$$

The reason for this choice will be presented in Section 3.5 from the viewpoint of linear programming.

3.4. Linear Programming Algorithm

We show that the completely ergodic Markovian decision process may also be formulated as a linear programming problem. For any stationary strategy of the completely ergodic process, the average return per unit time is

$$\mathfrak{g}(f) = \pi(f) r(f) \tag{3.37}$$

where $\pi(f)$ is the limiting vector of $P(f)$, which satisfies (3.18) through (3.20). Thus our problem is to find an optimal strategy f^* such that

$$\pi(f^*) r(f^*) = \max_{f \in F} \pi(f) r(f) \tag{3.38}$$

It is convenient to extend the range of decisions to include randomized strategies. Note that the assumption of complete ergodicity also holds for the extension to randomized strategies. Then, let d_j^k $(j \in S, k \in K_j)$ be the joint probability that the system is in state i and the decision k is made, where d_j^k is independent of time n because we restrict ourselves to stationary strategies. It is evident that

$$\sum_{k \in K_j} d_j^k = 1, \qquad 0 \leqslant d_j^k \ (\leqslant 1) \qquad \text{for} \quad j \in S, \quad k \in K_j \tag{3.39}$$

Thus the objective function of our problem is

$$\sum_{j \in S} \sum_{k \in K_j} \pi_j(f) r_j^k d_j^k \tag{3.40}$$

from (3.38) and the definition of d_j^k. The constraints using d_j^k are

$$\pi_j(f) - \sum_{i \in S} \sum_{k \in K_j} \pi_i(f) p_{ij}^k d_i^k = 0 \qquad \text{for} \quad j \in S \tag{3.41}$$

$$\sum_{j \in S} \pi_j(f) = 1 \tag{3.42}$$

$$\pi_j(f) > 0 \qquad \text{for} \quad j \in S \tag{3.43}$$

Setting

$$x_j^k = \pi_j(f)d_j^k \geqslant 0 \tag{3.44}$$

and using the fact that $\pi_j(f) = \sum_{k \in K_j} x_j^k$ for $j \in S$, we have the following

linear programming problem:

$$\max \sum_{j \in S} \sum_{k \in K_j} r_j^k x_j^k \tag{3.45}$$

subject to

$$\sum_{k \in K_j} x_j^k - \sum_{i \in S} \sum_{k \in K_i} p_{ij}^k x_i^k = 0 \qquad \text{for} \quad j \in S \tag{3.46}$$

$$\sum_{j \in S} \sum_{k \in K_j} x_j^k = 1 \tag{3.47}$$

$$x_j^k \geqslant 0 \qquad \text{for} \quad j \in S, \quad k \in K_j \tag{3.48}$$

For the completely ergodic process, rank $(I - P) = N - 1$, since rank $P^* = 1$. Thus, one of the constraints (3.46) is redundant. We omit one constraint, for example, for $j = N$ of (3.46). Then the constraints are (3.46) for $j = 1, 2, \ldots, N - 1$, and (3.47).

THEOREM 3.8. *There esists a basic feasible solution with the property that, for each $i \in S$, there is only one k such that $x_i^k > 0$, and $x_i^k = 0$ for k otherwise.*

PROOF. Our linear programming problem has N constraints, where one redundant constraint is omitted. Since the rank of the constraints is N, there are N positive variables x_i^k with the others zero for any basic feasible solution as follows (from the basic properties of linear programming). For the coefficients of the constraints, $-p_{ij}^k$ $(i \neq j)$ is nonpositive, $(1 - p_{ii}^k)$ is positive, and x_i^k is nonnegative; hence, there is at least one term $(1 - p_{ii}^k)$ in which x_i^k is positive for each i. That is, for each i there is at least one $x_i^k > 0$. If there are two $x_j^k > 0$ for any one j, there is some i without the term $(1 - p_{ii}^k)$ somewhere, since any basic feasible solution has N positive variables x_j^k with the others zero. This contradicts the fact that the right-hand side is zero or unity. Therefore, for each i, there is only one $x_j^k > 0$ with the others zero. ∎

COROLLARY 3.9. *Any basic feasible solution of the linear programming problem* (3.45) *through* (3.48) *yields a pure stationary strategy.*

PROOF. From (3.44) and $\pi_j(f) = \sum_{k \in K_j} x_j^k$, we have

$$d_j^k = x_j^k \Big/ \sum_{k \in K_j} x_j^k \tag{3.49}$$

Thus $d_j^k = 0$ or 1 from Theorem 3.8. ∎

From the above discussion, an optimal solution of the linear programming problem [(3.45) through (3.48)] yields an *optimal pure stationary strategy* f^∞. Also the primal variable $x_j^k > 0$ gives the limiting probability $\pi_j(f) > 0$ for state j.

3.5. Relationship between the Two Algorithms

We have shown in Section 2.4 that policy iteration and linear programming algorithms are equivalent in mathematical programming for the discounted Markovian decision process. In this section we also demonstrate the similar relationship for the completely ergodic process.

We now rewrite a linear programming problem discussed in the preceding section as a *primal problem*.

Primal problem:

$$\max \sum_{j \in S} \sum_{k \in K_j} r_j^k x_j^k \tag{3.50}$$

subject to

$$\sum_{k \in K_j} x_j^k - \sum_{i \in S} \sum_{k \in K_i} p_{ij}^k x_i^k = 0 \qquad \text{for} \quad j = 1, 2, \ldots, N-1 \tag{3.51}$$

$$\sum_{j \in S} \sum_{k \in K_j} x_j^k = 1 \tag{3.52}$$

$$x_j^k \geqslant 0 \qquad \text{for} \quad j \in S, \quad k \in K_j \tag{3.53}$$

Here we omit a redundant constraint for $j = N$ in (3.51).

The *dual problem* of the above is significant because any basic feasible solution corresponds to the N dual equalities (which are nonsingular) from Theorem 3.8 and its proof. Let the dual variables be $(v_1, v_2, \ldots, v_N)$.

Dual problem:

$$\min v_N \tag{3.54}$$

subject to

$$v_N + v_i \geqslant r_i^k + \sum_{j=1}^{N-1} p_{ij}^k v_j \qquad \text{for} \quad i \in S, \quad k \in K_i \tag{3.55}$$

$$v_i, \text{ unconstrained in sign} \qquad \text{for} \quad i \in S \tag{3.56}$$

Since the primal problem has an optimal solution and its value is g, we may write the dual problem using the *duality theorem* (see, for example, Dantzig [25]) as follows.

Dual problem:

$$\max g \tag{3.57}$$

subject to

$$g + v_i \geqslant r_i^k + \sum_{j=1}^{N-1} p_{ij}^k v_j \qquad \text{for} \quad i \in S, \quad k \in K_i \tag{3.58}$$

$$g, v_i, \text{ unconstrained in sign} \qquad \text{for} \quad i = 1, 2, \ldots, N - 1 \tag{3.59}$$

where we suppose that $v_N = 0$ in (3.58).

The dual problem is immediately derived from the discussion of Section 3.3. Since the method of deriving the dual problem is almost similar to that of Section 2.2, we omit the derivation here.

It is more comprehensive to write the Tucker diagram (Table 3.1) for the primal and dual problems. In the dual problem, setting $v_N = 0$ corresponds to omitting a redundant constraint for $j = N$ in (3.51) of the primal problem.

Since the discussion is similar to that of Section 2.4, we can omit phase I and can immediately obtain a basic feasible solution using Theorem 3.8. As an initial basic variable, we may, for example, apply

$$-r_j^* = \min_{k \in K_j} \left[-r_j^k \right] \qquad \text{for} \quad j \in S \tag{3.60}$$

For any basic feasible solution, we have the basic matrix

$$B = \begin{bmatrix} 1 - p_{11}^* & -p_{21}^* & \cdots & -p_{N_1}^* \\ -p_{12}^* & 1 - p_{22}^* & \cdots & -p_{N_2}^* \\ \cdot & \cdot & & \cdot \\ \cdot & \cdot & & \cdot \\ \cdot & \cdot & & \cdot \\ -p_{1,N-1}^* & -p_{2,N-1}^* & \cdots & -p_{N,N-1}^* \\ 1 & 1 & \cdots & 1 \end{bmatrix} \tag{3.61}$$

where p_{ji}^* is the transition probability corresponding to a basic solution. From the similar discussion of Section 2.4, defining the *simplex multipliers* (also *dual variables*) $\mu = (v_1, v_2, \ldots, v_{N-1}, g)$, we have

$$B^T \mu^T = r^* \tag{3.62}$$

where r^* is the $N \times 1$ column vector whose ith element is r_i^*, and the superscript T denotes the transpose. Rewriting (3.62) element by element, and supposing $v_N = 0$, we have

$$g + v_i = r_i^* + \sum_{j=1}^{N-1} p_{ij}^* v_j \qquad \text{for} \quad i \in S \tag{3.63}$$

which corresponds to the value determination operation in the policy iteration algorithm discussed in Section 3.3.

The simplex criterion of the next step is

$$[\Delta_i^k] = \left[-r_i^k - \sum_{j=1}^{N-1} p_{ij}^k v_j + g + v_i \right] \tag{3.64}$$

TABLE 3.1

The Tucker Diagram for the Completely Ergodic Markovian Decision Process

		Primal									
	Variables	$x_1^1 \geqslant 0$	$x_1^2 \geqslant 0$	$\cdots$	$x_2^1 \geqslant 0$	$x_2^2 \geqslant 0$	$\cdots$	$x_N^1 \geqslant 0$	$x_N^2 \geqslant 0$	$\cdots$	Relations	Constants
Dual	v_1	$1-p_{11}^1$	$1-p_{11}^2$	$\cdots$	$-p_{21}^1$	$-p_{21}^2$	$\cdots$	$-p_{N1}^1$	$-p_{N1}^2$	$\cdots$	$=$	0
	v_2	$-p_{12}^1$	$-p_{12}^2$		$1-p_{22}^1$	$1-p_{22}^2$	$\cdots$	$-p_{N2}^1$	$-p_{N2}^2$	$\cdots$	$=$	0
	$\cdots$	$\cdots$	$\cdots$		$\cdots$	$\cdots$		$\cdots$	$\cdots$		$\cdots$	$\cdots$
	v_{N-1}	$-p_{1,N-1}^1$	$-p_{1,N-1}^2$	$\cdots$	$-p_{2,N-1}^1$	$-p_{2,N-1}^2$	$\cdots$	$-p_{N,N-1}^1$	$-p_{N,N-1}^2$	$\cdots$	$=$	0
	g	1	1		1	1	$\cdots$	1	1	$\cdots$	$=$	1
	Relations	$\wedge$	$\wedge$		$\wedge$	$\wedge$	$\cdots$	$\wedge$	$\wedge$	$\cdots$		
	Constants	r_1^1	r_1^2		r_2^1	r_2^2	$\cdots$	r_N^1	r_N^2	$\cdots$		

It is evident that for the basic variables, we have

$$\Delta_i^* = - r_i^* - \sum_{j=1}^{N-1} p_{ij}^* v_j + \mathfrak{g} + v_i = 0 \qquad \text{for} \quad i \in S \qquad (3.65)$$

If for all $i \in S$ and $k \in K_i$,

$$\Delta_i^k = - r_i^k - \sum_{j=1}^{N-1} p_{ij}^k v_j + \mathfrak{g} + v_i \geqslant 0 \qquad (3.66)$$

or using (3.65), for all $i \in S$ and $k \in K_i$

$$r_i^* + \sum_{j=1}^{N-1} p_{ij}^* v_j \geqslant r_i^k + \sum_{j=1}^{N-1} p_{ij}^k v_j \qquad (3.67)$$

we have an optimal solution from the basic theory of linear programming.

If there is at least one pair $i \in S$ and $k \in K_i$ such that

$$\Delta_i^k = - r_i^k - \sum_{j=1}^{N-1} p_{ij}^k v_j + \mathfrak{g} + v_i < 0 \qquad (3.68)$$

or

$$r_i^* + \sum_{j=1}^{N-1} p_{ij}^* v_j < r_i^k + \sum_{j=1}^{N-1} p_{ij}^k v_j \qquad (3.69)$$

there exists an improved solution, or an improved strategy, that corresponds to the *policy improvement routine*. From the viewpoint of linear programming, we may apply $f(i) = k$ such that

$$\max_{k \in K_i} \left[r_i^k + \sum_{j=1}^{N-1} p_{ij}^k v_j \right] \qquad (> r_i^* + \sum p_{ij}^* v_j) \qquad (3.70)$$

We have already seen from Theorem 3.8 that substitutions for many variables imply an improved strategy.

Consequently, we have the equivalence between policy iteration and linear programming algorithms. That is, the policy iteration algorithm is a special extension of linear programming, such that pivot operations for many (at most N) variables are performed simultaneously.

3.6. Examples

The examples described in Section 2.6 are completely ergodic. In this section we solve two examples, the taxicab problem and the automobile replacement problem, by using the policy iteration and linear programming algorithms. Further, we propose a new algorithm, which is a mixture of the above two algorithms.

Taxicab problem. (Howard [63, p. 44].) The data of the problem were given in Table 2.2. Starting with an initial strategy, we have an optimal strategy with three iterations. The calculations are summarized in Table 3.2. We also show the linear programming solution starting with the same initial strategy (that is, the initial basic feasible solution) of the policy iteration algorithm. Then we get an optimal solution with six steps, where we suppose that three steps are required to get an initial basic feasible solution.

TABLE 3.2

Solution of the Taxicab Problem by the Policy Iteration Algorithm

v_1	1.33		-3.88		-1.18			
v_2	7.47		12.85		12.66			
v_3	0		0		0			
g	9.20		13.15		13.34			
	↗ ↘		↗ ↘		↗ ↘			
	VDO[a]	PIV[b]	VDO	PIV	VDO	PIV		
	1		1		2		2	
f	1		2		2		2	STOP
	1		2		2		2	

[a] VDO is the value determination operation.
[b] PIV is the policy improvement routine.

From R. A. Howard, "Dynamic Programming and Markov Processes." M.I.T. Press, Cambridge, Massachusetts, 1960.

Now we consider a new algorithm that is a mixture of the above two algorithms. We recall the relationship between the two algorithms. As we pointed out in Section 2.4, the policy iteration algorithm has a drawback as far as computation is concerned. That is, even if there is only one pair $i \in S$ and $k \in K_i$ satisfying (3.69), we must solve the system of N linear equations. However, the standard linear programming approach also has a disadvantage, especially, for large-scale problems because the simplex criteria are always required in each step. But the basic feasible solution may be changed for one variable in each step by using the standard linear programming code. From Theorems 3.7 and 3.8, we have seen that an improved strategy makes the increase of the average return g. In other words, the pivot operations for many variables yield an improved strategy. Thus we come to a new algorithm. We apply the linear programming approach, except that the simplex criteria

4

are used in the same fashion as the policy iteration algorithm. The new algorithm is summarized as follows:

1. Take an initial strategy and calculate an initial basic feasible solution by using Theorem 3.8.

2. Calculate

$$\Delta_i^k = - r_i^k - \sum_{j=1}^{N-1} p_{ij}^k v_j + g + v_i$$

for each $i \in S$ and $k \in K_i$, where we suppose $v_N = 0$.

3. If $\Delta_i^k \geqslant 0$ for all $i \in S$ and $k \in K_i$, we get an optimal solution. Or, if $\Delta_i^k < 0$ for some $i \in S$ and $k \in K_i$, perform pivot operations (change the basic variables) for each $i \in S$ and its corresponding $k \in K_i$ which satisfies $\Delta_i^k < 0$, by using the simplex tableau; then return to Step 2.

This new algorithm can be applied to the discounted Markovian decision model in Chapter 2 by performing the suitable modification. We omit the algorithm here.

Figure 3.3 shows the results of the three algorithms, where we assume that one iteration corresponds to 3 ($= N$) steps of linear programming. Figure 3.1 shows that the new algorithm is the most efficient among the three.

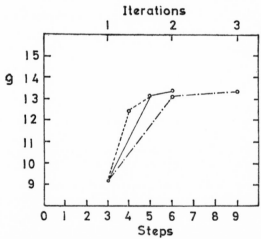

FIG. 3.1. Comparison among three algorithms for the taxicab problem. Policy iteration algorithm —.—.—, new algorithm ———, linear programming - - - - -.

In general, the new algorithm is more efficient than the policy iteration one because calculating pivot operations such that $\Delta_i^k < 0$ for each $i \in S$ and its associated $k \in K_i$ is easier than solving the system of N linear equations. We cannot, however, answer whether the new algorithm or the linear programming algorithm is more efficient.

Automobile replacement problem. (Howard [63, p. 54].) Let us consider the problem of automobile replacement over a time interval of ten years. We agree to review our current situation every three months and to make a decision whether to keep our present car or to trade it in at that time. The state of the system, i, is described by the age of the car in three-month periods; i may run from 1 to 40. In order to keep the number of states finite, a car of age 40 remains a car at age 40 forever (it is considered to be essentially worn out). The actions available in each state are these: The first action, $k = 1$, is to keep the present car for another quarter. The other actions, $k > 1$, are to buy a car of age $k = 2$, where $k - 2$ may be as large as 39. We then have 40 states with 41 actions in each state, with the result that there are 41^{40} possible stationary strategies.

The data supplied are the following: C_i is the cost of buying a car of age i; T_i, the trade-in value of a car of age i; E_i, the expected cost of operating a car of age i until it reaches age $i + 1$; and p_i, the probability that a car of age i will survive to be $i + 1$ without incurring a prohibitively expensive repair.

The probability defined here is necessary to limit the number of states. A car of any age that has a hopeless breakdown is immediately sent to state 40. Naturally, $p_{40} = 0$.

The data r_i^k and p_{ij}^k, by using the terms of our earlier notation, can be written as

$$r_i^k = - E_i \quad \text{for} \quad k = 1$$

$$r_i^k = T_i - C_{k-2} - E_{k-2} \quad \text{for} \quad k > 1$$

$$p_{ij}^k = \begin{cases} p_i & j = i + 1 \\ 1 - p_i & j = 40 \\ 0 & \text{other} \end{cases} \quad \text{for} \quad k = 1$$

$$p_{ij}^k = \begin{cases} p_{k-2} & j = i + 1 \\ 1 - p_{k-2} & j = 40 \\ 0 & \text{other} \end{cases} \quad \text{for} \quad k > 1$$

The actual data used in the problem are listed in Table 3.3 and graphed in Fig. 3.2. The discontinuities in the cost and trade-in functions were introduced in order to characterize typical model-year effects.

The policy iteration algorithm yields an optimal strategy with seven iterations. The linear programming algorithm yields an optimal strategy with

TABLE 3.3

Data for the Automobile Replacement Problem

Age in periods i	Cost C_i	Trade-in value T_i	Operating expense E_i	Survival probability P_i	Age in periods i	Cost C_i	Trade-in value T_i	Operating expense E_i	Survival probability P_i
0	$2000	$1600	$50	1.000	21	$345	$240	$115	0.925
1	1840	1460	53	0.999	22	330	225	118	0.919
2	1680	1340	56	0.998	23	315	210	121	0.910
3	1560	1230	59	0.997	24	300	200	125	0.900
4	1300	1050	62	0.996	25	290	190	129	0.890
5	1220	980	65	0.994	26	280	180	133	0.880
6	1150	910	68	0.991	27	265	170	137	0.865
7	1080	840	71	0.988	28	250	160	141	0.850
8	900	710	75	0.985	29	240	150	145	0.820
9	840	650	78	0.983	30	230	145	150	0.790
10	780	600	81	0.980	31	220	140	155	0.760
11	730	550	84	0.975	32	210	135	160	0.730
12	600	480	87	0.970	33	200	130	167	0.660
13	560	430	90	0.965	34	190	120	175	0.590
14	520	390	93	0.960	35	180	115	182	0.510
15	480	360	96	0.955	36	170	110	190	0.430
16	440	330	100	0.950	37	160	105	205	0.300
17	420	310	103	0.945	38	150	95	220	0.200
18	400	290	106	0.940	39	140	87	235	0.100
19	380	270	109	0.935	40	130	80	250	0
20	360	255	112	0.930					

R. A. Howard, "Dynamic Programming and Markov Processes." M.I.T. Press, Cambridge, Massachusetts, 1960.

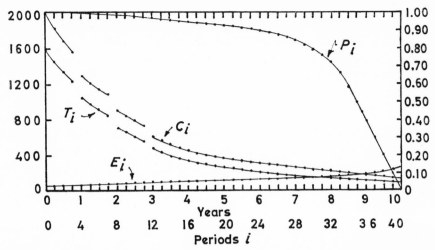

FIG. 3.2. Graphs for the automobile replacement data. (From R. A. Howard, "Dynamic Programming and Markov Processes." M.I.T. Press, Cambridge, Massachusetts, 1960.)

84 steps, where we suppose that the calculation of finishing phase I (obtaining a basic feasible solution) needs 40 ($= N$) steps. The new algorithm yields an optimal strategy with 174 steps. Here we apply the same strategy as the initial one. Figure 3.3 shows the results of the above three algorithms.

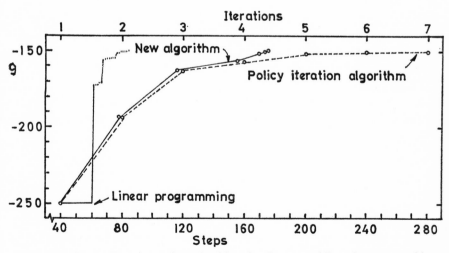

FIG. 3.3. Comparison among three algorithms for the automobile replacement problem.

The new algorithm is more efficient than the policy iteration one. That is, the new algorithm requires the same size simplex criteria, but for the pivot operations we can save $\frac{174}{280}$ the computing time (about 62%). Comparison with linear programming is generally an open question.

3.7. Terminating Processes

In this section we discuss a Markovian decision process of special type. That is, we assume that the system has a common absorbing state, regardless of the decisions we make. Here we denote the absorbing state by state 1. Then we suppose that state 1 is reachable from any other state. This assumption is described precisely as follows:

Terminating assumption. The common absorbing state can be reached with probability 1 from any state in a finite number of transitions, regardless of the decisions made.

In other words, this assumption asserts that state 1 is absorbing, and states $2, \ldots, N$ are transient regardless of the decisions made.

Now our problem is to find a sequence of decisions, namely a strategy, maximizing the total expected return before absorption, and the associated maximum value. This problem appeared in Shapley's [105] stochastic game in which the second player is a dummy (see appendix); in general the problem is a terminating stochastic game. Shapley's terminating assumption is $\sum_{j=2}^{N} p_{ij}^k < 1$ for all i and k. Our assumption is at least $\sum_{j=2}^{N} p_{ij}^k < 1$ for some i and all its associated actions $k \in K_i$. The system may not move to state 1 in a transition from some transient state. But state 1 is reachable from any transient state in a finite number of transitions with probability 1.

The objective function of our problem is finite because the system is absorbed in a finite number of transitions with probability 1. This is an example in which the total expected return is finite for the nondiscounted model.

The total expected return starting in each state, using any strategy π is

$$V(\pi) = \sum_{i=0}^{\infty} P_i(\pi) r(f_{i+1}) \tag{3.71}$$

where $\pi = (f_1, f_2, \ldots, f_i, \ldots)$, $V(\pi)$, $r(f_i)$ are $(N - 1) \times 1$ column vectors and $P_i(\pi)$ is the $(N - 1) \times (N - 1)$ matrix obtained by eliminating the first row and column. For this problem, we have the following theorems. The proof is almost similar to that of Section 2.2. We therefore omit the proof.

LEMMA 3.10. $V(\pi^*) \geqslant V(f, \pi^*)$ *for all* $f \in F$ *implies that* π^* *is optimal, where* (f, π^*) *is a strategy preceding* π^* *with* f.

THEOREM 3.11. *Exactly one of the following has to occur for each* $g \in F$:
(a) $V(f^\infty) \geqslant V(g, f^\infty)$ *for all* $g \in F$ *implies that* f^∞ *is optimal.*

(b) $$V(f^\infty) < V(g, f^\infty)$$
for some $g \in F$ *implies that*
$$V(f^\infty) < V(g, f^\infty) < V(g^\infty)$$

THEOREM 3.12. *There is an optimal strategy that is stationary.*

Theorem 3.11 gives the following *policy iteration algorithm:*

Value Determination Operation

Take any $f \in F$. Solve
$$v_i = r_i^k + \sum_{j=2}^{N} p_{ij}^k v_j$$
for v_i $(i = 2, \ldots, N)$, where the superscript k corresponds to the chosen strategy f^∞.

Policy Improvement Routine

Using the values v_i $(i = 2, \ldots, N)$, find the element of $G(i, f)$ for each $i = 2, \ldots, N$ such that
$$r_i^k + \sum_{j=2}^{N} p_{ij}^k v_j > v_i$$
for all $k \in K_i$. If $G(i, f)$ is empty for all $i = 2, \ldots, N$, f^∞ is optimal and $V(f^\infty) = [v_i]$ is the total expected return. If $g(i) \in G(i, f)$ for some i, use an improved strategy g^∞ such that $g(i) \in G(i, f)$ for some i and $g(i) = f(i)$ for $G(i, f)$ empty; then return to the value determination operation.

We also have the following *linear programming* problem for the terminating process:
$$\max \sum_{j=2}^{N} \sum_{k \in K_j} r_j^k x_j^k \tag{3.72}$$
subject to
$$\sum_{k \in K_j} x_j^k - \sum_{i=2}^{N} \sum_{k \in K_i} p_{ij}^k x_i^k = a_j \quad \text{for} \quad j = 2, \ldots, N \tag{3.73}$$
$$x_j^k \geqslant 0 \quad \text{for} \quad j = 2, \ldots, N; \quad k \in K_j \tag{3.74}$$
where $a = (a_1, a_2, \ldots, a_N)$ is an initial distribution for the process.

This derivation of the above linear programming problem is almost similar to that of Section 2.3. Thus the same properties discussed in Section 2.3 are also satisfied. Further we can readily show the relation between the above two algorithms. We do not present the details here.

The similarity between the discounted process and the terminating process will be shown in Chapter 8 using the principle of contraction mappings.

References and Comments

We treat only finite Markov chains. Here the classification of states obeys that of Kemeny and Snell [70]. Some properties of Markov chains are found in many texts of Markov chains, for example, Kemeny and Snell [70]. Summability of sequences is found in, for example, Kemeny et al. [71].

The policy iteration algorithm for the completely ergodic process has been given by Howard [63]. Optimality of stationary strategies has been given by Derman [36], Blackwell [14], and Martin-Löf [87]. Linear programming formulations have been given by De Ghellinck [27], Manne [85], Derman [36], and Wolfe and Dantzig [120]. Comparison among the three algorithms has been given by Osaki (to appear in Keieì-Kagaku).

The terminating process has been discussed by Howard [63, appendix], Eaton and Zadeh [53], and Osaki and Mine [95].

For the completely ergodic process, it is important to select an initial strategy. The other successive approximations have also been studied. White [116] has given a successive approximation method using the approximation in policy space of dynamic programming. Norman and White [92] have given a method for approximating an optimal strategy by using expectations.

Chapter 4

Markovian Decision Processes with no Discounting II

4.1. Introduction

In the preceding chapter we discussed special structure problems of Markovian decision processes with no discounting, that is, the completely ergodic process and the terminating process. In this chapter we discuss the general model with no discounting.

For the nondiscounted models, Denardo and Fox [33] classify four cases as follows:

1. The completely ergodic case—all states belong to a single ergodic set under every strategy.

2. The simple single chain case—all strategies produce a single ergodic set plus a (perhaps empty) strategy-dependent set of transient states.

3. The general single chain case—an optimal strategy with a single ergodic set and a (possibly empty) set of transient states.

4. The multichain case—several ergodic sets and some transient states, which may vary from strategy to strategy.

We considered Case 1 in the preceding section. Cases 2 and 3 will be treated by modifying Case 1 in Section 4.4. Here we focus on the Case 4. Since there exists an optimal stationary strategy according to the average return criterion from Theorem 3.4, we can discuss only stationary strategies. In Sections 4.2 and 4.3, our approach for the nondiscounted model treats $\beta = 1$ as a limiting case of $\beta < 1$, since we can apply the known facts about the discounted model. In this case we encounter a problem for finding 1-optimal strategies (definition below), where 1-optimal strategies coincide with strategies that maximize not only the average return but also the bias term. In Section 4.2 we give the policy iteration algorithm for finding an optimal strategy under the average return criterion, and we discuss the properties of 1-optimal strategies. In Section 4.3 we establish the policy iteration algorithm for finding a 1-optimal strategy. In Section 4.4 we present a linear programming problem for the general case and show that its dual corresponds to the policy iteration algorithm.

4.2. Policy Iteration Algorithm

We treat $\beta = 1$ as a limiting case of $\beta < 1$. For the discounted model, the discounted total expected return is

$$V_\beta(\pi) = \sum_{i=0}^{\infty} \beta^i P_i(\pi) r(f_{i+1})$$

where $0 \leqslant \beta < 1$. Let $\pi(\beta)$ be β-optimal and $U(\beta) = V_\beta(\pi(\beta))$.

DEFINITION 4.1. *A strategy π^* is called* 1-optimal *if*

$$\lim_{\beta \to 1-0} [V_\beta(\pi^*) - U(\beta)] = 0 \tag{4.1}$$

It will be shown that 1-optimal strategies are important for our problem. We now show the following theorem, which corresponds to Lemma 3.6.

THEOREM 4.2. *For any stationary strategy f^∞, let*

$$P^*(f) = \lim_{n \to \infty} \sum_{i=0}^{n-1} P^i(f)/n$$

which has been defined in Lemma 3.3. Then

$$V_\beta(f^\infty) = u(f)/(1 - \beta) + v(f) + \varepsilon(\beta, f) \tag{4.2}$$

where $u(f)$ is a unique solution of

$$(I - P(f))u = 0, \qquad P^*(f)u = P^*(f)r(f) \tag{4.3}$$

$v(f)$ is a unique solution of

$$(I - P(f))v = r(f) - u(f), \qquad P^*(f)v = 0 \tag{4.4}$$

and $\varepsilon(\beta, f) \to 0$ as $\beta \to 1 - 0$.

PROOF.   We have

$$V_\beta(f^\infty) = \sum_{i=0}^{\infty} \beta^i P^i(f) r(f)$$

$$= \left[\sum_{i=0}^{\infty} \beta^i P^*(f) + \sum_{i=0}^{\infty} \beta^i (P^i(f) - P^*(f)) \right] r(f)$$

$$= \frac{P^*(f)r(f)}{1 - \beta} + H(f)r(f) + [H(\beta, f) - H(f)]r(f) \tag{4.5}$$

where $H(f)$ and $H(\beta, f)$ are defined in (3.4). Thus we have (4.2) by setting $u(f) = P^*(f)r(f)$, $v(f) = H(f)r(f)$, and $\varepsilon(\beta, f) = [H(\beta, f) - H(f)]r(f)$. It is clear from Lemma 3.3 that $u(f)$ and $v(f)$ are unique solutions of (4.3) and (4.4), respectively. ∎

The solutions $u(f)$ and $v(f)$ of (4.3) and (4.4) may be written as

$$u(f) = P^*(f)r(f) \tag{4.6}$$

$$v(f) = H(f)r(f) = [(I - P(f) + P^*(f))^{-1} - P^*(f)]r(f) \tag{4.7}$$

The above equations state that, specifying $f \in F$, we can determine $P^*(f)$ and can also obtain $u(f)$ and $v(f)$. Thus our problem is to find a strategy that maximizes $u(f)$ [and also $v(f)$] within all stationary strategies. That is, our problem is a combinatorial one. We need the efficient algorithm for this problem.

As we have seen in Theorem 3.4, there exists an optimal stationary strategy under the average return criterion; thus we can define a set of strategies such that

$$F' = \{f \mid f \in F, u(f) \geqslant u(g) \quad \text{for all} \quad g \in F\} \tag{4.8}$$

That is, F' is the set of all $f \in F$ having maximal *average return per unit time*. Further, we can define a set of strategies such that

$$F'' = \{f \mid f \in F', v(f) \geqslant v(g) \quad \text{for all} \quad g \in F'\} \tag{4.9}$$

It is evident that F'' is a subset of F' and is the set of all $f \in F'$ having maximal *bias term $v(f)$*.

For the discounted model, $V_\beta(g, f^\infty)$ is computed by using (4.2) as follows:

$$V_\beta(g, f^\infty) = r(g) + \beta P(g)V_\beta(f^\infty)$$

$$= \beta P(g)u(f)/(1 - \beta) + r(g) + \beta P(g)v(f) + \beta P(g)\varepsilon(\beta, f)$$

$$= P(g)u(f)/(1 - \beta) + r(g) - P(g)u(f) + P(g)v(f)$$

$$\qquad - (1 - \beta)P(g)v(f) + \beta P(g)\varepsilon(\beta, f)$$

$$= P(g)u(f)/(1 - \beta) + r(g) - P(g)u(f) +$$

$$\qquad + P(g)v(f) + \varepsilon_1(\beta, f, g) \tag{4.10}$$

where

$$\varepsilon_1(\beta, f, g) = -(1 - \beta)P(g)v(f) + \beta P(g)\varepsilon(\beta, f) \to 0$$

as $\beta \to 1 - 0$.

Comparing the right-hand sides of both (4.2) and (4.10), we define the set of actions such that

$$G(i, f) = \{k \mid k \in K_i, \sum_{j \in S} p_{ij}^k u_j > u_i \quad \text{or} \quad \sum_{j \in S} p_{ij}^k u_j = u_i$$

$$\text{and} \quad r_i^k + \sum_{j \in S} p_{ij}^k v_j > u_i + v_i\} \tag{4.11}$$

where u_i and v_i are the ith elements of $u(f)$ and $v(f)$, respectively. Further we define

$$G(f) = \underset{i=1}{\overset{N}{\times}} G(i, f) \tag{4.12}$$

THEOREM 4.3. *Take any $f \in F$. (a) If $G(f)$ is empty, then $f \in F'$. (b) If $g(i) \in G(i, f)$ for some $i \in S$ and $g(i) = f(i)$ whenever $g(i) \notin G(i, f)$, then $V_\beta(g^\infty) > V_\beta(f^\infty)$ for all β (< 1) sufficiently near 1.*

PROOF.   If $G(f)$ is empty, that is, if $G(i, f)$ is empty for all $i \in S$,

$$u(f) \geqslant P(g)u(f), \qquad \text{or} \quad u(f) = P(g)u(f)$$

and

$$v(f) > r(g) - P(g)u(f) + P(g)v(f) = r(g) - u(f) + P(g)v(f)$$

Thus, we have $V_\beta(g, f^\infty) \leqslant V_\beta(f^\infty)$ for all β sufficiently near 1. From Theorem 2.3, $V_\beta(f^\infty) \geqslant V_\beta(g^\infty)$ for all $g \in F$ and all β sufficiently near 1. Thus, from the expression (4.2) we have $f \in F'$.

   Proof of (b). If $g(i) \in G(i, f)$ for some i and $g(i) = f(i)$ whenever $g(i) \notin G(i, f)$, then $V_\beta(g, f^\infty) > V_\beta(f^\infty)$ for all β sufficiently near 1. From Theorem 2.4 we have $V_\beta(g^\infty) > V_\beta(f^\infty)$ for all β sufficiently near 1. ∎

This theorem describes the policy iteration algorithm, which yields an optimal stationary strategy $f \in F'$ with finite iterations.

COROLLARY 4.4.   *For any $f \in F$, take $g \in F$ such that $g(i) \in G(i, f)$ for some i and $g(i) = f(i)$ whenever $g(i) \notin G(i, f)$. Then either $u(g) > u(f)$, or $u(g) = u(f)$ and $v(g) > v(f)$.*

PROOF.   From (b) of Theorem 4.3 and the expression (4.2), we have $u(g) \geqslant u(f)$; and $u(g) = u(f)$ implies that $v(g) \geqslant v(f)$. It remains only to show that $(u(g), v(g)) \neq (u(f), v(f))$. If we assume $(u(g), v(g)) = (u(f), v(f))$ then $G(f)$ is empty, which is a contradiction. ∎

For each i, we define the set of actions such that

$$E(i, f) = \{k | k \in K_i, \sum_{j \in S} p_{ij}^k u_j = u_i \qquad \text{and}$$

$$r_i^k + \sum_{j \in S} p_{ij}^k v_j = u_i + v_i\} \tag{4.13}$$

which depends only on the data p_{ij}^k, r_i^k for the actions of state $i \in S$. Further we define

$$E(f) = \underset{i=1}{\overset{N}{\times}} E(i, f) \tag{4.14}$$

It is evident that $E(f)$ contains at least f.

LEMMA 4.5.   *If $f \in F$ and $g \in E(f)$, then $u(g) = u(f)$. If in addition $P^*(g)P^*(f) = P^*(g)$, then $v(g) = v(f)$.*

PROOF. $g \in E(f)$ is equivalent to, writing u, v for $u(f), v(f)$,

$$P(g)u = u \tag{4.15}$$

and

$$r(g) + P(g)v = u + v \tag{4.16}$$

Premultiplying (4.16) by $P^*(g)$ yields

$$P^*(g)r(g) = P^*(g)u \tag{4.17}$$

But from Theorem 4.2, Eqs. (4.15) and (4.17) have a unique solution $u = u(g)$, so that $u(g) = u(f)$. Also from $P^*(f)v = 0$, we have $P^*(g)P^*(f)v = 0$, so that, if $P^*(g)P^*(f) = P^*(g)$, we have

$$P^*(g)v = 0 \tag{4.18}$$

But since $u = u(g)$, a unique solution of (4.16) and (4.18) is $v = v(g)$, so that $v(g) = v(f)$. ∎

THEOREM 4.6.   *If $G(f)$ is empty and $g \in E(f)$ implies*

$$P^*(g)P^*(f) = P^*(g) \tag{4.19}$$

then f is 1-optimal.

PROOF.   Let f satisfy the hypothesis of the theorem, and choose β near enough to 1 that, for any pair f_1, f_2, we have $V_\beta(f_1, f_2^\infty) \geqslant V_\beta(f_2^\infty)$, which implies that $f_2(i) \in G(i, f_1) \cup E(i, f_1)$ for all $i \in S$. If our f is not β-optimal, let $f_0 = f_1, f_2, \ldots, f_k$ be a sequence of β-improvements, obtained as in Theorem 2.5, terminating in a β-optimal f_k. Then

$$f_{n+1}(i) \in G(i, f_n) \cup E(i, f_n) \tag{4.20}$$

for all $n = 0, 1, \ldots, k - 1$. We show by induction on n that $u(f_n) = u(f_0)$ and $v(f_n) = v(f_0)$. This is true for $n = 0$. If true for a given n, then, since $G(i, f)$ and $E(i, f)$ depend only on $u(f)$ and $v(f)$, $G(i, f_n)$ is empty and $E(i, f_n) = E(i, f)$. That is, $f_{n+1}(i) \in E(i, f_0)$ for all $n = 0, 1, \ldots, k - 1$. Then f, f_{k+1} satisfy the hypothesis of f, g in Lemma 4.5, so that $u(f_{k+1}) = u(f)$, $v(f_{k+1}) = v(f)$. Thus, writing $f(\beta)$ for the β-optimal f_k, we have

$$U(\beta) = u(f)/(1 - \beta) + v(f) + \varepsilon(\beta, f_\beta) \tag{4.21}$$

Since

$$V_\beta(f^\infty) = u(f)/(1 - \beta) + v(f) + \varepsilon(\beta, f) \tag{4.22}$$

we have $U(\beta) - V_\beta(f^\infty) \to 0$ as $\beta \to 1 - 0$, and f^∞ is 1-optimal. ∎

COROLLARY 4.7.   *If $G(f)$ is empty, and $E(f)$ contains only f, then f is 1-optimal.*

PROOF.   For any $f \in F$,
$$P^*(f)P^*(f) = P^*(f) \tag{4.23}$$
which corresponds to (4.19). From Theorem 4.6 we have a 1-optimal f. ∎

The following theorem gives the important property of 1-optimal strategies.

THEOREM 4.8. F'' is the nonempty set of all $f \in F$ for which f^∞ is 1-optimal.

PROOF.   If $G(f)$ is empty, we have, from (4.10), the inequality
$$V_\beta(g, f_0^\infty) \leqslant V_\beta(f_0^\infty) + \tau(\beta)\mathbf{1} \tag{4.24}$$
for β near 1, where $\tau(\beta)$ is a scalar function of β, the maximum element of $\varepsilon_1(\beta, f_0, g) - \varepsilon(\beta, f_0)$. We have $\tau(\beta) \to 0$ as $\beta \to 1 - 0$. Denoting $L(g)$ in Section 2.2 by L, we write (4.24) as $LV_\beta(f_0) \leqslant V_\beta(f_0) + \tau(\beta)\mathbf{1}$ for near 1. We show by induction on n that for all n,
$$L^n V_\beta(f_0) \leqslant V_\beta(f_0) + (1 + \beta + \cdots + \beta^{n-1})\tau(\beta)\mathbf{1} \tag{4.25}$$
for β near 1. If (4.25) holds for a given n, we obtain, applying L,
$$
\begin{aligned}
L^{n+1} V_\beta(f_0) &\leqslant L[\text{right-hand side of (4.25)}] \\
&= r(g) + \beta P(g)V_\beta(f_0) + \beta(1 + \beta + \cdots + \beta^{n-1})\tau(\beta)\mathbf{1} \\
&= V_\beta(g, f_0^\infty) + \beta(1 + \beta + \cdots + \beta^{n-1})\tau(\beta)\mathbf{1} \\
&\leqslant V_\beta(f_0) + (1 + \beta + \cdots + \beta^n)\tau(\beta)\mathbf{1}
\end{aligned}
\tag{4.26}
$$
where the last inequality is obtained by using (4.24).

Thus
$$L^n V_\beta(f_0) \leqslant V_\beta(f_0) + [\tau(\beta)/(1 - \beta)]\mathbf{1}$$
for all n, so that, for all $g \in F$,
$$V_\beta(g) = \lim_{n \to \infty} L^n V_\beta(f_0) \leqslant V_\beta(f_0) + [\tau(\beta)/(1 - \beta)]\mathbf{1} \tag{4.27}$$
for β near 1. But
$$
\begin{aligned}
V_\beta(g) - V_\beta(f_0) &= [u(g) - u(f_0)]/(1 - \beta) + v(g) \\
&\quad - v(f_0) + \varepsilon(\beta, g) - \varepsilon(\beta, f_0)
\end{aligned}
\tag{4.28}
$$
(4.27) and (4.28) implies $u(g) \leqslant u(f_0)$.

Take any f^* which is β-optimal for a set of β's having 1 as a limit point. From (4.28), with $g = f^*$ we obtain $u(f^*) \geqslant u(f_0)$, so that $u(f^*) = u(f_0)$. For any $g \in F'$, we have
$$V_\beta(f^*) - V_\beta(g) = v(f^*) - v(g) + \varepsilon(\beta, f^*) - \varepsilon(\beta, g) \tag{4.29}$$
so that, letting $\beta \to 1 - 0$ through a sequence for which f^* is β-optimal, we obtain $v(f^*) \geqslant v(g)$ for all $g \in F'$. Thus $f^* \in F''$. ∎

This theorem states that 1-*optimal strategies* have not only the *maximal average return* $u(f)$ but also the *maximal bias term* $v(f)$ in $f \in F$. Thus if there are two or more strategies with the same average return, we shall require a 1-optimal strategy. But for the case of Corollary 4.7, we have immediately a 1-optimal strategy. In general we fail to find 1-optimal strategies. Consider two simple numerical examples (Blackwell [14]).

EXAMPLE 1.   There are two states, 1 and 2, and two actions, 1 and 2. In state 1, action 1 yields the return of 1, and the system remains in state 1 with probability $\frac{1}{2}$ and moves to state 2 with probability $\frac{1}{2}$, while action 2 yields the return of 2 and the system moves to state 2 with certainty. In state 2, either action yields the return of 0 and the system remains in state 2. Now $F = \{f, g\}$, where $f(1) = 1, g(1) = 2, f(2) = g(2) = 1$. Then

$$P(f) = \begin{bmatrix} \frac{1}{2} & \frac{1}{2} \\ 0 & 1 \end{bmatrix}, \qquad r(f) = \begin{bmatrix} 1 \\ 0 \end{bmatrix}$$

$$P(g) = \begin{bmatrix} 0 & 1 \\ 0 & 1 \end{bmatrix}, \qquad r(g) = \begin{bmatrix} 2 \\ 0 \end{bmatrix}$$

Thus we have

$$V_\beta(f) = \begin{bmatrix} 2/(2 - \beta) \\ 0 \end{bmatrix}, \quad V_\beta(g) = \begin{bmatrix} 2 \\ 0 \end{bmatrix}$$

So, $U(\beta) = 2$ and f^∞ is 1-optimal. It is clear that f satisfies the hypothesis of Lemma 4.5.

EXAMPLE 2.   There are also two states, 1 and 2, and two actions. In state 1, action 1 yields 3 and the system remains in state 1 with probability $\frac{1}{2}$. Action 2 yields 6, and the system moves to state 2 with certainty. In state 2, either action loses 3 and the system remains in state 2 with probability $\frac{1}{2}$ and moves to state 1 with probability $\frac{1}{2}$. Now $F = \{f, g\}$, where $f(1) = 1, g(1) = 2$, and $f(2) = g(2) = 1$. Then

$$P(f) = \begin{bmatrix} \frac{1}{2} & \frac{1}{2} \\ \frac{1}{2} & \frac{1}{2} \end{bmatrix}, \qquad r(f) = \begin{bmatrix} 3 \\ -3 \end{bmatrix}$$

$$P(g) = \begin{bmatrix} 0 & 1 \\ \frac{1}{2} & \frac{1}{2} \end{bmatrix}, \qquad r(g) = \begin{bmatrix} 6 \\ -3 \end{bmatrix}$$

Thus we have

$$u(f) = u(g) = \begin{bmatrix} 0 \\ 0 \end{bmatrix}, \qquad v(f) = \begin{bmatrix} 3 \\ -3 \end{bmatrix}, \qquad v(g) = \begin{bmatrix} 4 \\ -2 \end{bmatrix}$$

so that

$$V_\beta(g) - V_\beta(f) \rightarrow \begin{bmatrix} 1 \\ 1 \end{bmatrix}$$

as $\beta \to 1 - 0$ and f is not 1-optimal. It is evident that g is 1-optimal. If we start with an initial strategy f, we fail to find a 1-optimal strategy, as we see from Theorem 4.3. But if we start with g, we will find 1-optimal strategy.

As we have seen the above examples, we cannot always find a 1-optimal strategy by using Theorem 4.3. Hence we need an algorithm for finding 1-optimal strategies. Section 4.3 will present a computing algorithm for finding a 1-optimal strategy.

In the rest of this section, we characterize the set inclusion of F' and $E(f)$ and discuss other properties.

LEMMA 4.9. *For any $f, g \in F$, the Markov chain defined by $P(g)$ has no transient states:* (a) *If $g \in G(f)$, then $u(g) > u(f)$.* (b) *If $G(f)$ is empty and $g \notin E(f)$, then $u(f) > u(g)$.*

PROOF.   From Corollary 4.4, $g \in G(f)$ implies that $u(f) \leqslant u(g)$. From (a) of Theorem 4.3, $G(f)$ empty implies that $u(f) \geqslant u(g)$. Thus it suffices to show that $u(f) \neq u(g)$ in each case.

Suppose to the contrary that $u(f) = u(g)$. Then by (3.3)

$$P^*(g)[P(g)u(f) - u(f)] = P^*(g)u(f) - P^*(g)u(f) = 0 \qquad (4.30)$$

But, from the hypothesis that there are no transient states, $P^*(g) \geqslant 0$ has a positive element in each column; and $P(g)u(f) - u(f)$ is nonnegative if $g \in G(f)$, and nonpositive if $G(f)$ is empty. Thus

$$P(g)u(f) = u(f) \qquad (4.31)$$

Also by (4.6) and (3.3), we have

$$P^*(g)[r(g) + P(g)v(f) - u(f) - v(f)]$$
$$= u(g) + P^*(g)v(f) - u(g) - P^*(g)v(f) = 0 \qquad (4.32)$$

Since $P^*(g)$ has a positive element in each column, and since by (4.31) the bracketed term in (4.32) is nonnegative if $g \in G(f)$ and nonpositive if $G(f)$ is empty, we have

$$r(g) + P(g)v(f) = u(f) + v(f) \qquad (4.33)$$

It follows from this equation and (4.31) that $g \in E(f)$. This contradicts the hypothesis $g \in G(f)$ in (a) and $g \notin E(f)$ in (b). ∎

Lemma 4.9 states that the policy iteration algorithm is more efficient under the hypothesis of the lemma. That is, if $P(g)$ has no transient states, we have always $u(g) > u(f)$ for $g \in G(f)$, or $u(f) > u(g)$ if $G(f)$ is empty and $g \notin E(f)$.

COROLLARY 4.10.   *For any $f \in F'$, we have the following:* (a) *$E(f) \subset F'$.*

(b) *If the Markov chain defined by $P(g)$ has no transient states for each $g \in F$, then $G(f)$ is empty and $E(f) = F'$.*

PROOF. That of (a) is straightforward from Lemma 4.5.

Proof of (b). Suppose first that $g \in G(f)$. Then $u(f) < u(g)$ by (a) of Lemma 4.9, which contradicts the hypothesis that $f \in F'$. Thus $G(f)$ is empty. Now suppose that $g \in F' - E(f)$. Then by (b) of Lemma 4.9, $u(g) < u(f)$ which contradicts the hypothesis that $g \in F'$. Thus $F' \subset E(f)$. Combining this fact with (a) completes the proof. ∎

If the Markov chain defined by $P(g)$ has no transient states for each $g \in F$, then (a) of Theorem 4.3 and (b) of Corollary 4.8 hold, and the policy iteration algorithm terminates immediately upon finding an element of F'. Moreover, $E(f)$ is precisely the set of all g with the maximal average return per unit time. Thus, if the algorithm is initiated with $f \in F' - F''$, the algorithm does not yield a 1-optimal strategy. In Example 2 we encountered this case.

LEMMA 4.11. *If $f, g \in F''$, then $E(f) = E(g)$. Moreover, $F'' \subset E(f) \subset F'$.*

PROOF. The set $E(f)$ depends only on $(u(f), v(f))$, and similarly for g. But $f, g \in F''$ implies that $(u(f), v(f)) = (u(g), v(g))$; hence $E(f) = E(g)$. Also since $g \in E(g) = E(f)$, $F'' \subset E(f)$. Finally, $E(f) \subset F'$ follows from Corollary 4.10. ∎

EXAMPLE 3. There are two states, 1 and 2, and five actions in state 2. In state 1, every action yields 2 and the system moves to state 2 with certainty. In state 2, action k ($k = 1, 2, 3, 4, 5$) yields $3 + k/5$ and the system remains in state 2 with probability $k/5$ and moves to state 1 with probability $(5 - k)/5$. Now $F = \{f_1, f_2, f_3, f_4, f_5\}$, where $f_k(1) = 1, f_k(2) = k$ ($k = 1, 2, 3, 4, 5$). Then

$$P(f_k) = \begin{bmatrix} 0 & 1 \\ k/5 & (5-k)/5 \end{bmatrix}, \qquad r(f_k) = \begin{bmatrix} 2 \\ 3 + k/5 \end{bmatrix}$$

we have

$$F'' = \{f_5\} \neq F' = \{f_1, f_2, f_3, f_4, f_5\} = F$$

which is an example that F'' is a proper subset of F'. Then f_5 is 1-optimal.

4.3. Policy Iteration Algorithm for Finding 1-Optimal Strategies

In this section we shall consider an algorithm for finding a 1-optimal strategy. As we have seen in the preceding section, we have found an $f' \in F'$ with $G(f')$ empty by employing the policy iteration algorithm in Theorem 4.3.

5

From Theorem 4.8 we can find an $f'' \in F''$ over all $f' \in F'$. For the special case when there are no transient states in each $f \in F$, we can characterize all $f' \in F' = E(f')$ from (b) of Corollary 4.10. But in general we cannot characterize all elements of the set F'. We can apply the fact that $E(f')$ is immediately available, and that, from Corollary 4.10, we have $E(f') \subset F'$. This suggests that we consider instead maximizing $v(g)$ over $g \in E(f')$, $f' \in F'$. It turns out that this problem can be solved with the techniques already developed. We introduce a new (vector) quantity $w(f)$ by means of the following lemma.

LEMMA 4.12. *For any $f \in F$ and $g \in E(f)$, let $w(g)$ be a unique solution of*

$$[I - P(g)]w = 0, \qquad P^*(g)w = P^*(g)(-v(f)) \tag{4.34}$$

Then

$$v(g) = v(f) + w(g) \tag{4.35}$$

PROOF. The uniqueness of $w(g)$ follows from Theorem 4.2. Since $g \in E(f)$, we have

$$[I - P(g)]v(f) = r(g) - u(f) \tag{4.36}$$

Adding this equation to the first equation in (4.34) and then rewriting the second equation in (4.34), we obtain

$$[I - P(g)][v(f) + w(g)] = r(g) - u(f), \qquad P^*(g)[v(f) + w(g)] = 0 \tag{4.37}$$

But from Theorem 4.3 a unique solution of this system is $v(g)$. Thus (4.35) holds. ∎

Note from (4.3), (4.14), and Lemma 4.12 that the problem of maximizing $w(g)$ over $g \in E(f)$ has the same form as that of maximizing $u(g)$ over $g \in F$, where we replace K_i by $E(i, f)$, F by $E(f)$, and $r(g)$ by $-v(f)$ for $i \in S$ and all $g \in E(f)$. Thus the policy iteration algorithm of Theorem 4.3 can be used to find a g that maximizes $w(g)$—and hence $v(g)$ in view of (4.35)—over $g \in E(f)$.

As we have discussed in the first part of this section, if we find $f' \in F'$ with $G(f')$ empty then we may find an $f'' \in E(f')$ that maximizes $v(g)$ over $g \in E(f')$. But in general we know that $E(f')$ may be a proper subset of F' in which case we need an algorithm for determining if $f'' \in F''$. An algorithm of this type is given in the next lemma.

LEMMA 4.13. *For an $f \in F$, if $G(f)$ is empty, and if $v(f) \geq v(g)$ for all $g \in E(f)$ then $f \in F''$.*

PROOF. The proof is a slight modification of Theorem 4.6. Choose $\beta < 1$ near 1 so that for any f_0, f_1 with $G(f_0)$ empty, we have $V_\beta(f_1, f_0^\infty) \geq V_\beta(f_0^\infty)$, which implies that $f_1 \in E(f_0)$, and $V_\beta(f_1) \geq V_\beta(f_0)$ and $u(f_1) = u(f_0)$, which imply that $v(f_1) \geq v(f_0)$. If f^∞ is not β-optimal, let $f_0 = f$ and let $f_1, f_2, \ldots, f_k$ be a sequence of β-improvements, obtained as in Theorem 2.5, terminating in a β-optimal f_k^∞. We show by induction on n that $(u(f_n), v(f_n)) = (u(f_0), v(f_0))$, and $G(f_n)$ is empty for $n = 0, 1, \ldots, k$. This is true for $n = 0$. If true for a given n, then $E(f_n) = E(f)$ because $E(f)$ depends only on $(u(f), v(f))$. Thus $f_{n+1} \in E(f)$, so by Lemma 4.5, $u(f_{n+1}) = u(f)$. Moreover, $V_\beta(f_{i+1}) > V_\beta(f^\infty)$, so by the definition of β, $v(f_{n+1}) \geq v(f)$. But by hypothesis, $v(f_{n+1}) \leq v(f)$, so necessarily $v(f_{n+1}) = v(f)$. Since $(u(f_{n+1}), v(f_{n+1})) = (u(f_0), v(f_0))$ and $G(f_0)$ is empty, $G(f_{i+1})$ is also empty.

Consequently, writing $f(\beta)$ for the β-optimal f_k, we have, from (4.2), that

$$U(\beta) = u(f)/(1 - \beta) + v(f) + \varepsilon(\beta, f(\beta)) \tag{4.38}$$

and

$$V_\beta(f^\infty) = u(f)/(1 - \beta) + v(f) + \varepsilon(\beta, f) \tag{4.39}$$

so $U(\beta) - V_\beta(f^\infty) \to 0$ as $\beta \to 1 - 0$. Thus f^∞ is 1-optimal and so by Theorem 4.8, $f \in F''$. ∎

The next corollary slightly generalizes Theorem 4.6.

COROLLARY 4.14.   *If $f \in F$ and if $G(f)$ is empty, and if $P^*(g)P^*(f) = P^*(g)$ for all $g \in E(f)$, then $E(f) = F''$.*

PROOF.   By (a) of Theorem 4.3 and Lemma 4.5, $(u(f), v(f)) = (u(g), v(g))$ for all $g \in E(f)$. Thus $E(g) = E(f)$ and $G(g) = G(f)$ for $g \in E(f)$, so by Lemma 4.13, $E(f) \subset F''$. On the other hand, by Lemma 4.11, $F'' \subset E(f)$. ∎

For each $f \in F$ let $z(f)$ be the unique solution (by Theorem 4.2) of

$$[I - P(f)]z(f) = -v(f), \qquad P^*(g)z(f) = 0 \tag{4.40}$$

In the same way as we have defined the set $G(i, f)$, we define a set of actions such that for any f,

$$\Theta(i, f) = \{k | k \in E(i, f), -v_i + \sum_{j \in S} p_{ij}^k z_j > z_i\} \tag{4.41}$$

where v_i and z_i are the ith element of $v(f)$ and $z(f)$, respectively. We also define

$$\Theta(f) = \mathop{\times}_{i=1}^{N} \Theta(i, f) \tag{4.42}$$

where if $\Theta(i, f)$ is empty for all $i \in S$, we call $\Theta(f)$ empty; and if $g(i) \in \Theta(i, f)$ for some i, $g \in \Theta(f)$ means that $g(i) \in \Theta(i, f)$ and $g(i) = f(i) \notin \Theta(i, f)$ for each $i \in S$.

Using the sets $G(f)$ and $\Theta(f)$, we have the *policy iteration algorithm for finding a 1-optimal strategy*. We now summarize the policy iteration algorithm, which is our main result in this section.

THEOREM 4.15. *Take any $f \in F$.* (a) *If $G(f)$ is empty, then $f \in F'$.* (b) *If $G(f) \cup \Theta(f)$ is empty, then $f \in F''$.* (c) *If $g \in G(f)$, then either $u(g) > u(f)$, or $u(g) = v(g)$ and $v(f) > v(g)$.* (d) *If $g \in \Theta(f)$, then $u(g) = u(f)$; and either $v(g) > v(f)$, or $v(g) = v(f)$ and $z(g) > z(f)$.*

PROOF.   The proofs of (a) and (c) follow from Theorem 4.3 and Corollary 4.4.

Proof of (b).   We first note that $w(f) = 0$ from (4.35). Since $\Theta(f)$ is empty, it follows from Theorem 4.3 [if $\Theta(f)$ is empty, then $0 = w(f) \geqslant w(g)$ for all $g \in E(f)$] and Lemma 4.12 [$v(g) = v(f) + w(g)$] that

$$v(f) = v(f) + w(f) = v(g) - w(g) \geqslant v(g)$$

for all $g \in E(f)$. Thus $f \in F''$.

Proof of (d).   It is evident that $g \in \Theta(f) \subset E(f)$, so we have from Lemma 4.5 that $u(g) = u(f)$. It follows from Theorem 4.3 (if $g \in \Theta(f)$, then $V_\beta(g^\infty) \geqslant V_\beta(f^\infty)$ for all β sufficiently near 1), from Lemma 4.12, and from part (c) above, that either $w(g) > w(f) = 0 [v(g) > v(f)]$, or $w(g) = w(f) = 0$ $[v(g) = v(f)]$ and $z(g) > z(f)$ from (4.35). ∎

This theorem gives the policy iteration algorithm for finding a 1-optimal strategy. That is, specify an initial strategy $f_1 \in F$, and choose $f_2, f_3, \ldots$, inductively to satisfy $f_{n+1} \in G(f_n) \cup \Theta(f_n)$. The sequence $\{f_n\}$ is finite because F is finite, and the sequence of triples $\{u(f_n), v(f_n), z(f_n)\}$ is *lexicographically increasing* so that no f_n can recur. Thus, $G(f_n) \cup \Theta(f_n)$ must be empty for some n. Here, "lexicographically smaller" means that, for any two same-dimension vectors u and v, if $u \neq v$ and the first nonzero component of $v - u$ is positive, then u is lexicographically smaller than v.

In practice we may apply the policy iteration algorithm using $G(f)$ and after $G(f)$ empty, we may again apply the policy iteration algorithm using $\Theta(f)$. We can show the procedure for the policy iteration algorithm for finding a 1-optimal strategy as follows.

1. Take any $f \in F$.

2. Compute $u(f)$ and $v(f)$. If $G(f)$ is empty, then $f \in F'$; then go to Step 2. Otherwise, make an improved strategy $g \in G(f)$ and repeat the first part of this step.

3. Specify $E(f)$ and compute $z(f)$. If $\Theta(f)$ is empty, then $f \in F''$. Otherwise make an improved strategy $g \in \Theta(f)$ and repeat the first part of this step.

Now we shall demonstrate the policy iteration algorithm for finding a 1-optimal strategy for Example 3 in Section 4.2. The data of Example 3 are given on page 53.

As an initial strategy, we take f_1. Then we have

$$u(f_1) = \begin{bmatrix} 3 \\ 3 \end{bmatrix}, \qquad v(f_1) = \begin{bmatrix} -\frac{5}{6} \\ \frac{1}{6} \end{bmatrix}$$

Using $u(f_1)$ and $v(f_1)$, we know that $G(f_1)$ is empty, and thus $f_1 \in F'$. Then we go to Step 3 of the above procedure.

Specifying $E(f_1)$, we have $E(f_1) = \{f_1, f_2, f_3, f_4, f_5\}$. Computing $z(f_1)$, we have

$$z(f_1) = \begin{bmatrix} \frac{25}{36} \\ -\frac{5}{36} \end{bmatrix}$$

Using $v(f_1)$ and $z(f_1)$, we have

$$-v_2 + \sum_{j=1}^{2} p_{ij}^k z_j = -\frac{11}{36} + \frac{k}{6}$$

Thus we have an improved strategy f_5 that maximizes

$$(-v_2 + \sum_{j=1}^{2} p_{ij}^k z_j > z_2).$$

Computing $v(f_5)$ and $z(f_5)$ again, we know that $\Theta(f_5)$ is empty, which yields $f_5 \in F''$.

Note that Example 3 is a special case where $P(f)$ has no transient states for each $f \in F$. From Corollary 4.10, the emptiness of $G(f)$ implies that $E(f) = F'$.

4.4. Linear Programming Algorithm

In this section we develop a linear programming algorithm for the general Markovian decision process with no discounting. In this section we apply an approach of treating directly the long-run average return per unit time. Thus our problem is to find a strategy that maximizes the average return per unit time. From Theorem 3.4, we restrict our attention to *stationary strategies* and write a stationary strategy f instead of f^∞.

From (3.12) we have, for any $f \in F$,

$$a\Gamma(f) = aP^*(f)r(f) = \sum_{i \in S} \sum_{j \in S} a_i p_{ij}^*(f) r_j(f) \tag{4.43}$$

which is the average return per unit time starting in an initial distribution a, where $P^*(f) = [p_{ij}^*(f)]$. Note that an optimal strategy under the average return criterion is independent of the initial distribution since an optimal strategy is attained simultaneously for each initial state. Thus we consider a problem for finding a stationary strategy that maximizes (4.43) under all $f \in F$. In other words, our problem is to find an optimal strategy f that yields

$$\max_{f \in F} \sum_{i \in S} \sum_{j \in S} a_i p_{ij}^*(f) r_j(f) \qquad (4.44)$$

It is convenient to extend the range of decisions to include randomized strategies. Let d_j^k denote the joint probability that the system is in state $j \in S$ and that an action $k \in K_j$ is made. It is evident that

$$\sum_{k \in K_j} d_j^k = 1, \qquad d_j^k \geqslant 0 \qquad \text{for} \quad j \in S, \quad k \in K_j \qquad (4.45)$$

We still consider any fixed nonrandomized strategy later.

Setting

$$x_j^k = \sum_{i \in S} a_i p_{ij}^*(f) d_j^k \qquad \text{for} \quad j \in S, \quad k \in K_j \qquad (4.46)$$

and

$$y_j^k = \sum_{i \in S} a_i h_{ij}(f) d_j^k \qquad \text{for} \quad j \in S, \quad k \in K_j \qquad (4.47)$$

where

$$[h_{ij}(f)] = H(f) = (I - P(f) + P^*(f))^{-1} - P^*(f) \qquad (4.48)$$

and using the relations $P^*(I - P^*) = 0$ [from (3.3)] and $P^* + H(I - P) = I$ [from (3.6)], we have

$$\sum_{i \in S} \sum_{k \in K_i} (\delta_{ij} - p_{ij}^k) x_i^k = 0 \qquad \text{for} \quad j \in S \qquad (4.49)$$

and

$$\sum_{k \in K_j} x_j^k + \sum_{i \in S} \sum_{k \in K_i} (\delta_{ij} - p_{ij}^k) y_i^k = a_j \qquad \text{for} \quad j \in S \qquad (4.50)$$

where δ_{ij} is the Kronecker delta. It is evident from (2.3), (4.45), and (4.46) that

$$x_j^k = \sum_{i \in S} a_i p_{ij}^*(f) d_j^k \geqslant 0 \qquad \text{for} \quad j \in S, \quad k \in K_j \qquad (4.51)$$

The sign of y_j^k is not clear and we shall show later that $y_j^k \geqslant 0$ for any transient state.

Thus we have the following *linear programming* problem:

$$\max \sum_{j \in S} \sum_{k \in K_j} r_j^k x_j^k \qquad (4.52)$$

subject to

$$\sum_{i \in S} \sum_{k \in K_i} (\delta_{ij} - p_{ij}^k) x_i^k = 0 \qquad \text{for} \quad j \in S \qquad (4.53)$$

$$x_j^k \geqslant 0 \qquad \text{for} \quad j \in S, \quad k \in K_j \qquad (4.54)$$

$$\sum_{k \in K_j} x_j^k + \sum_{i \in S} \sum_{k \in K_i} (\delta_{ij} - p_{ij}^k) y_i^k = a_j \qquad \text{for} \quad j \in S \qquad (4.55)$$

We shall subsequently show that $y_j^k \geqslant 0$ for any transient state j.

For a fixed strategy f, Markov matrix $P(f)$ has some ergodic sets and a transient set. Appropriately relabeling the number of states, we have the following form for Markov matrix $P(f)$:

$$
P(f) = \begin{bmatrix}
P_{11} & & & & \text{\Large O} \\
& P_{22} & & & \\
& & \cdot & & \text{\Large O} \\
& & & \cdot & \\
\text{\Large O} & & & & \cdot & \\
& & & & & P_{vv} \\
\hline
P_{v+1,1} & P_{v+1,2} & \cdots & & P_{v+1,v} & P_{v+1,v+1}
\end{bmatrix} \tag{4.56}
$$

where $P_{11}, \ldots, P_{vv}$ are submatrices associated with each *ergodic set* E_μ ($\mu = 1, \ldots, v$), respectively, and the remaining states specify a set T of *transient states*.

The next two lemmas are useful for eliminating redundant constraints in (4.53) and (4.55). In the discussion of these lemmas (containing Lemma 4.18), we have to restrict ourselves to any fixed nonrandomized strategy, since we cannot consider simultaneously state classifications for randomized strategies. Thus we omit the summation on k.

LEMMA 4.16.   *Take any* $f \in F$. *For any* $j \in E_\mu$ ($\mu = 1, \ldots, v$), *constraints* (4.55) *become*

$$
\sum_{j \in E_\mu} x_j^k + \sum_{j \in E_\mu} \sum_{i \in T} (\delta_{ij} - p_{ij}^k) y_i^k = \sum_{j \in E_\mu} a_j \quad \text{for} \quad \mu = 1, \ldots, v \tag{4.57}
$$

PROOF.   Using part (c) of Lemma 3.3, we combine constraints (4.55) by summing on E_μ. From (4.56),

$$
\sum_{i \in S} = \sum_{i \in E_\mu} + \sum_{\substack{v=1 \\ v \neq \mu}} \sum_{i \in E_v} + \sum_{i \in T}
$$

and

$$
\sum_{i \in E\mu} (\delta_{ij} - p_{ij}^k) = \sum_{\substack{i \in E_\mu \\ v \neq \mu}} (\delta_{ij} - p_{ij}^k) = 0
$$

which imply (4.57). ∎

LEMMA 4.17.   *For any ergodic set* E_μ ($\mu = 1, \ldots, v$), *one of the constraints* (4.53) *associated with* E_μ *is redundant.*

PROOF.   The proof is obvious from the property of Markov chains and the fact that $P_{\mu\mu}$ ($\mu = 1, \ldots, v$) is a Markov matrix. ∎

Lemmas 4.16 and 4.17 give the necessary constraints for any *ergodic set*. The next lemma gives a constraint for any *transient state*.

LEMMA 4.18.   *Take any $f \in F$. For any state $j \in T$, constraint* (4.55) *becomes*

$$\sum_{i \in T} (\delta_{ij} - p_{ij}^k) y_i^k = a_j \qquad \text{for} \quad j \in T \tag{4.58}$$

where

$$y_i^k \geqslant 0 \qquad \text{for} \quad i \in T \tag{4.59}$$

PROOF.   From $[p_{ij}^*(f)]_{i \in S, j \in T} = [0]_{i \in S, j \in T}$ and (4.46), we have $x_i^k = 0$ for any $i \in T$ and $k = f(i)$. Also we have

$$\sum_{i \in S} (\delta_{ij} - p_{ij}^k) y_i^k = \sum_{i \notin T} (\delta_{ij} - p_{ij}^k) y_i^k + \sum_{i \in T} (\delta_{ij} - p_{ij}^k) y_i^k$$
$$= \sum_{i \in T} (\delta_{ij} - p_{ij}^k) y_i^k = a_j$$

which corresponds to (4.58). Furthermore, from

$$[h_{ij}(f)]_{i, j \in T} = [I - P_{v+1, v+1}]^{-1} = \sum_{n=0}^{\infty} P_{v+1, v+1}^n \geqslant 0,$$

$$[h_{ij}(f)]_{i \notin T, j \in T} = \left[\sum_{n=0}^{\infty} (P^n(f) - P^*(f))\right]_{i \notin T, j \in T} = [0]_{i \notin T, j \in T}$$

and (4.47), we have

$$y_i^k = \sum_{i \in S} a_i h_{ij}(f) d_i^k \geqslant 0 \qquad \text{for} \quad j \in T. \quad \blacksquare \tag{4.60}$$

From (4.59), we suppose that $y_j^k \geqslant 0$ for any $j \in S$ and $k \in K_j$, since y_j^k disappears for any ergodic state j.

Thus we also consider a dual problem of the linear programming problem (4.52), (4.53), (4.54), (4.55), and (4.59). Let the $N \times 1$ column vectors $u(f)$ and $v(f)$ be the corresponding dual variables. Then its *dual problem* is

$$\max \sum_{i \in S} a_i u_i \tag{4.61}$$

subject to

$$u_i \geqslant \sum_{j \in S} p_{ij}^k u_j \qquad \text{for} \quad i \in S, \quad k \in K_i \tag{4.62}$$

$$u_i + v_i \geqslant r_i^k + \sum_{j \in S} p_{ij}^k v_j \qquad \text{for} \quad i \in S, \quad k \in K_i \tag{4.63}$$

$$u_i, v_i; \qquad \text{unconstrained in sign for} \quad i \in S \tag{4.64}$$

where u_i, v_i are the ith elements of $u(f)$ and $v(f)$, respectively. We know that this dual problem corresponds to the policy iteration algorithm, that is, this dual problem is immediately derived from the policy iteration algorithm.

We also note that the dual variables $u(f)$ and $v(f)$ are unique solutions of

$$u(f) = P(f)u(f), \qquad u(f) + v(f) = r(f) + P(f)v(f) \qquad (4.65)$$

where from (3.7) we set the value of one v_i in each ergodic set to zero, which refers to (4.57). Then $v(f)$ is a *relative solution* and the difference between the exact solution of (4.4), and $v(f)$ in (4.65) is a constant.

We now have a linear programming algorithm for the general Markovian decision processes. The algorithm is constructed by using Lemmas 3.3 and 4.16 through 4.18. Further we note that the *dual variables* $(u(f), v(f))$ are also *simplex multipliers*. Using these simplex multipliers, we have the simplex criterion, which corresponds to policy improvement routine in the policy iteration algorithm. The direct proof of increasing the average return by its simplex criterion without linear programming properties can be made. It is convenient to consider the following set of actions, which corresponds to the *simplex criterion* of the primal problem, using simplex multipliers $[u(f), v(f)]$:

$$G(i, f) = \{k | k \in K_i, \sum_{j \in S} p_{ij}^k u_j > u_i \text{ or } \sum_{j \in S} p_{ij}^k u_j = u_i$$

$$\text{and} \quad r_i^k + \sum_{j \in S} p_{ij}^k v_j > u_i + v_i\} \qquad (4.66)$$

Then we have the following proposition describing the linear programming algorithm, which we give without proof.

PROPOSITION 4.19. *Taking any $f \in F$, determine the constraints for each state according to the state classification (use Lemmas 4.16 through 4.18); we can obtain a basic feasible solution that corresponds to a strategy f and that yields its dual variables $u(f)$ and $v(f)$ (simplex multipliers). Using these simplex multipliers, we have the simplex criterion. That is, if $G(i, f)$ is empty for all $i \in S$, then we have an optimal stationary strategy. Otherwise, we select a new strategy g such that $g(i) \in G(i, f)$ and $g(i) = f(i) \notin G(i, f)$. We then return to the first part of this proposition and repeat the procedures until an optimal strategy is obtained.*

Note that Proposition 4.19 describes a special structure linear programming algorithm such that pivot operations for many variables are performed simultaneously if there exist two or more states such that $G(i, f)$ is nonempty. Proposition 4.19 and the primal and dual problems imply the following corollary.

COROLLARY 4.20. *The policy iteration algorithm is equivalent to the linear programming one.*

PROOF. We must find a basic feasible solution (that is, the dual variables) that corresponds to value determination operation; then the simplex criterion of the next step corresponds to the policy improvement routine. But in the policy iteration algorithm, pivot operations are performed simultaneously for many variables. ∎

The next corollary is clear if we consider the policy iteration algorithm.

COROLLARY 4.21. *An optimal strategy is independent of the initial distribution a.*

We note from the above discussion that the policy iteration algorithm for the completely ergodic process can also be applied to Cases 2 and 3 in Section 4.1. That is, the policy iteration algorithm has the same form even if the process considered has a single ergodic set and some transient states, except that we set $v_i = 0$ for some ergodic state i.

Here we show the linear programming algorithm and the policy iteration algorithm for Howard's example [63, p. 65]. The data of the problem are given in Table 4.1. Let an initial strategy be denoted by f, its associated Markov matrix by $P(f)$, and the return vector by $r(f)$. Then

$$f = \begin{bmatrix} 3 \\ 2 \\ 1 \end{bmatrix}, \qquad P(f) = \begin{bmatrix} 0 & 0 & 1 \\ 0 & 1 & 0 \\ 1 & 0 & 0 \end{bmatrix}, \qquad r(f) = \begin{bmatrix} 3 \\ 4 \\ 8 \end{bmatrix}$$

The primal and dual problems are given in the following tableau (the *reduced Tucker diagram*), where only the data associated with a basic feasible solution are given. The superscripts for the strategy chosen are omitted.

$$
\begin{array}{c}
\quad\; x_1 \; x_2 \; x_3 \\
\begin{array}{c} v_1 \\ u_2 \\ u_1 = u_3 \end{array}
\begin{bmatrix} 1 & 0 & -1 \\ 0 & 1 & 0 \\ 1 & 0 & 1 \end{bmatrix}
\begin{array}{l} = 0 \\ = a_2 \\ = a_1 + a_3 \end{array}
\end{array}
\quad
\begin{array}{l} (v_2 = 0) \\ (v_3 = 0) \end{array}
\quad
u(f) = \begin{bmatrix} \frac{11}{2} \\ 4 \\ \frac{11}{2} \end{bmatrix}, \quad v(f) = \begin{bmatrix} -\frac{5}{2} \\ 0 \\ 0 \end{bmatrix}
$$

Dual variables

$$
\begin{array}{ccc}
\text{\Large\vee} & \text{\Large\vee} & \\
3 & 4 & 8
\end{array}
$$

Using the simplex criterion (or equivalently the policy improvement routine), we get an improved strategy g and its data:

$$g = \begin{bmatrix} 3 \\ 3 \\ 3 \end{bmatrix}, \qquad P(g) = \begin{bmatrix} 0 & 0 & 1 \\ 0 & 0 & 1 \\ 0 & 0 & 1 \end{bmatrix}, \qquad r(g) = \begin{bmatrix} 3 \\ 5 \\ 7 \end{bmatrix}$$

Then

$$
\begin{array}{c}
\begin{array}{cccc}
 & y_1 & y_2 & x_3 \\
\end{array} \\
\begin{array}{c}
v_1 \\
v_2 \\
u_1 = u_2 = u_3
\end{array}
\left|\begin{array}{ccc}
1 & 0 & 0 \\
0 & 1 & 0 \\
1 & 1 & 1
\end{array}\right|
\begin{array}{c}
= a_1 \\
= a_2 \\
= a_3 \quad (v_3 = 0),
\end{array} \\
\begin{array}{ccc}
\vee & \vee & \vee \\
3 & 5 & 7
\end{array}
\end{array}
\qquad
\text{Dual variables} \qquad
u(g) = \begin{bmatrix} 7 \\ 7 \\ 7 \end{bmatrix}, \qquad
v(g) = \begin{bmatrix} -4 \\ -2 \\ 0 \end{bmatrix}
$$

Using the simplex criterion, we get an optimal strategy g, since $G(i, g)$ is empty for any $i \in S$.

TABLE 4.1

Data for a Multichain Example

State	Action	Probability			Return
i	k	p_{i1}^k	p_{i2}^k	p_{i3}^k	r_i^k
1	1	1	0	0	1
	2	0	1	0	2
	3	0	0	1	3
2	1	1	0	0	6
	2	0	1	0	4
	3	0	0	1	5
3	1	1	0	0	8
	2	0	1	0	9
	3	0	0	1	7

From R. A. Howard, "Dynamic Programming and Markov Processes." M.I.T. Press, Cambridge, Massachusetts, 1960.

References and Comments

Four cases of the nondiscounted models in Section 9.1 have been considered by Denardo and Fox [33].†

The policy iteration algorithm for the general (or multichain) Markovian decision process was first given by Howard [63]. Blackwell [14] studied this problem rigorously and has pointed out the existence of 1-optimal strategies. Subsequently Veinott [111] gave an algorithm for finding 1-optimal strategies. The discussion in Sections 4.2 and 4.3 were first given by Blackwell [14] and Veinott [111].

A linear programming formulation of the general Markovian decision process has been given by Osaki and Mine [96].

Chapter 5

Dynamic Programming Viewpoint of Markovian Decision Processes

5.1. Introduction

In the preceding chapters we treated Markovian decision processes over an infinite time horizon. In this chapter we consider Markovian decision processes over a finite-time horizon. For the models with a finite-time horizon, *dynamic programming* is very useful. Thus in this chapter we consider the dynamic programming viewpoint of Markovian decision processes.

In Section 5.2, we give a dynamic programming formulation for Markovian decision processes over a finite-time horizon. An optimal policy (definition below) is then immediately derived by dynamic programming. We also study (in Section 5.3) the asymptotic behaviour of an optimal policy.

5.2. Dynamic Programming

Using the same notation that we used in the preceding chapters, we first define a policy. A *policy* π is a sequence $(\ldots, f_n, \ldots, f_1)$ of members of F, where f_n is the decisions for each state. That is, $f_n(i)$ is an action in state i measured n times backward from the end of the planning horizon. A policy π describes a backward sequence of actions in each state ending with f_1. The total expected return using n times of π starting in each state $i \in S$ is

$$v_n(\pi) = r(f_n) + P(f_n)r(f_{n-1}) + \cdots + P(f_n) \cdots P(f_2)r(f_1) \quad (5.1)$$

for $n \geqslant 1$, where $v_n(\pi)$ is the $N \times 1$ column vector whose ith element is the total expected return using n times of π starting in state $i \in S$. From (5.1) we have the following recurrence relation:

$$v_n(\pi) = r(f_n) + P(f_n)v_{n-1}(\pi) \quad (5.2)$$

for $n \geqslant 1$, where $v_0(\pi) = 0$.

Now we define optimal policies for $v_n(\pi)$.

DEFINITION 5.1. *A policy π is optimal if $v_n(\pi) \geqslant v_n(\sigma)$ for any policy σ and for all n.*

To find an optimal policy, we use the dynamic programming approach and make use of the principle of optimality.

Principle of optimality. An optimal policy has the property that despite the initial state and initial decisions, the remaining decisions must constitute an optimal policy for the state resulting from the first decision.

The principle of optimality is due to Bellman (see, for example, Bellman [7, p. 53]). Applying the principle of optimality, we have the following recurrence relation:

$$v_{n+1}(\pi^*)(i) = \max_{k \in K_i} \left[r_i^k + \sum_{j \in S} p_{ij}^k v_n(\pi^*)(j) \right] \tag{5.3}$$

for all $i \in S$ and for all $n \geqslant 0$, where

$$v_0(\pi^*)(i) = 0 \tag{5.4}$$

for all $i \in S$. Here $v_n(\pi^*)(i)$ is the ith element of $v_n(\pi^*)$. We can immediately obtain an optimal policy π^* from (5.3) and (5.4). An example is presented in Section 5.3.

We note that the dynamic programming approach can be applied to the case where r_i^k and p_{ij}^k depend on time n. The same recurrence relations [(5.3) and (5.4)] of dynamic programming can be applied to this case.

We have discussed the problem for finding an optimal policy over a finite-time horizon. Now we shall study an optimal policy over an infinite-time horizon. We assume then that r_i^k and p_{ij}^k are independent of time n.

We again use an operator $L(f)$ that maps X into $L(f)X = r(f) + P(f)X$, where X is any $N \times 1$ column vector. For any π and any $N \times 1$ column vector X, we define

$$
\begin{aligned}
V^n(\pi, X) &= L(f_n)L(f_{n-1}) \cdots L(f_1)X \\
&= r(f_n) + P(f_n)r(f_{n-1}) + \cdots + P(f_n) \cdots P(f_2)r(f_1) \\
&\qquad\qquad + P(f_n) \cdots P(f_1)X
\end{aligned} \tag{5.5}
$$

which is the total expected return vector starting in each state using n time of π, plus the extra n time expected return from X (the last term of the right-hand side). In particular, we have $V^n(\pi, 0) = v_n(\pi)$.

DEFINITION 5.2. *A policy π is X-optimal if $V^n(\pi, X) \geqslant V^n(\sigma, X)$ for any policy σ and for all $n \geqslant 1$.*

This definition implies that a policy π is *optimal* if it is *0-optimal*.

We shall now give three lemmas that will be useful in later discussion.

LEMMA 5.3. *For any $f \in F$,*

$$
\begin{aligned}
V_\beta(f^\infty) &= \sum_{i=0}^{\infty} \beta^i P^i(f)r(f) \\
&= [(1 - \beta)^{-1} - 1]u(f) + A(f) + \varepsilon(\beta, f)
\end{aligned} \tag{5.6}
$$

where $\varepsilon(\beta, f) \to 0$ as $\beta \to 1 - 0$ and $A(f) = [I - P(f) + P^*(f)]^{-1}r(f)$ (confer Theorem 4.2).

PROOF.   From Theorem 4.2, we have

$$V_\beta(f^\infty) = [P^*(f)r(f)]/(1 - \beta) + [I - P(f) + P^*(f)]^{-1}r(f)$$
$$- P^*(f)r(f) + \varepsilon(\beta, f) \qquad (5.7)$$

where $\varepsilon(\beta, f) \to 0$ as $\beta \to 1 - 0$. ∎

We define $L_\beta(f)X = r(f) + \beta P(f)X$, which was used in Section 2.2. In particular, $L_1(f)$ is abbreviated $L(f)$.

LEMMA 5.4

$$L(f)A(f) = u(f) + A(f) \qquad (5.8)$$

PROOF.   From (3.3), and the definition of $A(f)$, we have

$$L(f)A(f) = \lim_{\beta \to 1-0} [r(f) + \beta P(f)A(f)]$$

$$= \lim_{\beta \to 1-0} [r(f) + \beta P(f) \sum_{i=0}^{\infty} \beta^i(P(f) - P^*(f))^i r(f)]$$

$$= \lim_{\beta \to 1-0} [r(f) + \sum_{i=0}^{\infty} \beta^{i+1}(P(f) - P^*(f))^{i+1} r(f) + \beta P^*(f)r(f)]$$

$$= \lim_{\beta \to 1-0} [\sum_{i=0}^{\infty} \beta^i(P(f) - P^*(f))^i + \beta P^*(f)r(f)]$$

$$= A(f) + u(f). \qquad ∎ \qquad (5.9)$$

LEMMA 5.5.   *If f^∞ is an optimal strategy, then, for any $g \in F$, either*

(a) $P(g)u(f) < u(f)$ (5.10)

or

(b) $P(g)u(f) = u(f)$ and $L(g)A(f) \leqslant L(f)A(f)$ (5.11)

PROOF.   From Theorem 4.3, if f^∞ is an optimal strategy, then $P(g)u(f) < u(f)$, or $P(g)u(f) = u(f)$ and $r(g) + P(g)v(f) \leqslant u(f) + v(f)$ for any strategy g^∞. We have from (5.8):

$$L(f)A(f) = u(f) + A(f)$$

and from Theorem 4.2:

$$L(f)A(f) = u(f) + v(f) + u(f)$$

and

$$L(g)A(f) = r(g) + P(g)A(f)$$
$$= r(g) + P(g)v(f) + P(g)u(f) \qquad (5.12)$$

Thus, if $P(g)u(f) = u(f)$, then $r(g) + P(g)v(f) \leqslant u(f) + v(f)$ implies

$$r(g) + P(g)v(f) + u(f) \leqslant u(f) + v(f) + u(f)$$

that is, $L(g)A(f) \leqslant L(f)A(f)$. ∎

5.3. Properties of an Optimal Policy

We have seen in the preceding chapters that there is an optimal strategy that is stationary. For the problem to be considered now, is there an optimal policy that is stationary? An example follows.

EXAMPLE.   There are two states, 1 and 2, and two actions in state 1. In state 1, action 1 yields the return of 1, and the system remains in state 1 with probability $\frac{1}{2}$. Action 2 yields the return of $\frac{6}{5}$ and the system remains in state 1 with probability $\frac{1}{4}$. In state 2, either action yields the return of 0 and the system remains in state 2 with probability $\frac{1}{4}$. Now $F = \{f, g\}$, where $f(1) = 1$, $g(1) = 2$, and $f(2) = g(2) = 1$. Then

$$P(f) = \begin{bmatrix} \frac{1}{2} & \frac{1}{2} \\ \frac{3}{4} & \frac{1}{4} \end{bmatrix}, \qquad r(f) = \begin{bmatrix} 1 \\ 0 \end{bmatrix}$$

$$P(g) = \begin{bmatrix} \frac{1}{4} & \frac{3}{4} \\ \frac{3}{4} & \frac{1}{4} \end{bmatrix}, \qquad r(g) = \begin{bmatrix} \frac{6}{5} \\ 0 \end{bmatrix}$$

Thus we have

$$u(f) = u(g) = \begin{bmatrix} \frac{3}{5} \\ \frac{3}{5} \end{bmatrix}, \qquad A(f) = \begin{bmatrix} \frac{23}{25} \\ \frac{3}{25} \end{bmatrix}, \qquad A(g) = \begin{bmatrix} 1 \\ \frac{1}{5} \end{bmatrix}$$

and g^∞ is optimal (also 1-optimal—see Theorem 4.8).

Let us find the optimal policy. Let $X_n = \max_\pi V^n(\pi, 0)$, where $X_0 = 0$. Let us denote the optimal policy $\pi = (\ldots, f_n^*, \ldots, f_1^*)$. For any 2×1 column vector X_n, $L(f)X_n > L(g)X_n$ implies that $f_{n+1}^* = f$, or conversely $L(f)X_n \leqslant L(g)X_n$ implies that $f_{n+1}^* = g$. For any $x = \begin{bmatrix} x_1 \\ x_2 \end{bmatrix}$, we have

$$L(f)X = r(f) + P(f)X = \begin{bmatrix} 1 + \frac{1}{2}x_1 + \frac{1}{2}x_2 \\ \frac{3}{4}x_1 + \frac{1}{4}x_2 \end{bmatrix}$$

$$L(g)X = r(g) + P(g)X = \begin{bmatrix} \frac{6}{5} + \frac{1}{4}x_1 + \frac{3}{4}x_2 \\ \frac{3}{4}x_1 + \frac{1}{4}x_2 \end{bmatrix}$$

and $(1 + \frac{1}{2}x_1 + \frac{1}{2}x_2) - (\frac{6}{5} + \frac{1}{4}x_1 + \frac{3}{4}x_2) = \frac{1}{4}(x_1 - x_2 - \frac{4}{5})$. Thus we have that $x_1 - x_2 - \frac{4}{5} = T(X)R0$ if and only if $L(f)XRL(g)X$, where R is one of $>, =,$ or $<$. We have also $T(L(f)X) = -\frac{1}{4}T(X)$ and $T(L(g)X) = -\frac{1}{2}T(X)$. Hence the optimal policy is $\pi = (\ldots, f, g, f, g)$, an alternating sequence of f and g ending with g. That is, the optimal policy π is *periodic*.

To calculate the asymptotic return of a periodic policy, we only need to be able to calculate that of a stationary policy. For instance, in the above example, to calculate $V^{2n}(\pi, 0)$, we can define an operator $L(h) = L(f)L(g)$ and then calculate $L^n(h)(0)$. For a stationary policy, we have:

LEMMA 5.6.   *Denote $P(f)$ by P and $r(f)$ by r. Then: (a) If P is nonperiodic,*

$$V_n(f^\infty, X) = (n - 1)u(f) + A(f) + P^*X + \varepsilon(n, f) \qquad (5.13)$$

where $\varepsilon(n, f) \to 0$ as $n \to \infty$. (b) If P is periodic with period M, let $P_0 = P^M$, then

$$V^{nM+m}(f^\infty, X) = (n - 1)Mu(f) + [I - (P_0 - P_0^*)]^{-1}\Big[\sum_{i=0}^{m-1} P^i r\Big]$$
$$+ P_0^*\Big[\sum_{i=0}^{m-1} P^i r + P^m X\Big] + \varepsilon'(n, f) \qquad (5.14)$$

where $\varepsilon'(n, f) \to 0$ as $n \to \infty$, and $m = 0, 1, \ldots, M - 1; n = 0, 1, 2, \ldots.$

PROOF.   *Proof of (a).* Since P is nonperiodic, $\lim_{n \to \infty} P^n = P^*$. Thus, from Lemma 3.6, we have

$$V^n(f^\infty, X) = \sum_{i=0}^{n-1} P^i r + P^n X$$
$$= (n - 1)u(f) + A(f) + P^*X + \varepsilon(n, f) \qquad (5.15)$$

Proof of (b).   Using $P^M = P_0$, we have

$$V^{nM+m}(f^\infty, X) = \sum_{i=0}^{nM+m-1} P^i r + P^{nM+m}X$$
$$= (I + P_0 + \cdots + P_0^{n-1})\Big(\sum_{i=0}^{M-1} P^i r\Big) + P_0^n\Big(\sum_{i=0}^{m-1} P^i r + P^m X\Big)$$
$$= \Big[\sum_{i=0}^{n-1} (P_0 - P_0^*)^i\Big]\Big(\sum_{i=0}^{M-1} P^i r\Big) + (n - 1)P_0^*\Big(\sum_{i=0}^{M-1} P^i r\Big)$$
$$+ P_0^n\Big(\sum_{i=0}^{m-1} P^i r + P^m X\Big) \qquad (5.16)$$

Furthermore, $P_0^n \to P_0^*$ as $n \to \infty$, since P_0 is nonperiodic and

$$P^* = \lim_{n \to \infty} (nM + m)^{-1} \sum_{i=0}^{nM+m-1} P^i$$
$$= \lim_{n \to \infty} (nM + m)^{-1}\Big[\Big(\sum_{i=0}^{n-1} P_0^i\Big)\Big(\sum_{i=0}^{M-1} P^i\Big) + P_0^n\Big(\sum_{i=0}^{m-1} P^i\Big)\Big]$$
$$= \lim_{n \to \infty} n(nM + m)^{-1}\Big[\frac{1}{n}\Big(\sum_{i=0}^{n-1} P_0^i\Big)\Big(\sum_{i=0}^{M-1} P^i\Big) + \frac{P_0^n}{n}\Big(\sum_{i=0}^{m-1} P^i\Big)\Big]$$
$$= (M)^{-1}P_0^* \sum_{i=0}^{M-1} P^i \qquad (5.17)$$

We have $P_0^* \sum_{i=0}^{M-1} P^i = MP^*$. Hence we have

$$V^{nM+m}(f^\infty, X) = (n-1)Mu(f) + [I - (P_0 - P_0^*)]^{-1}\left[\sum_{i=0}^{m-1} P^i r\right]$$

$$+ P_0^*\left(\sum_{i=0}^{m-1} P^i r + P^m X\right) + \left[\sum_{i=n}^{\infty} (P_0 - P_0^*)^i\right]\left[\sum_{i=0}^{M-1} P^i X\right]$$

$$+ (P_0^n - P_0^*)\left(\sum_{i=0}^{m-1} P^i r + P^m X\right) \qquad (5.18)$$

setting

$$\varepsilon'(n, f) = \left[\sum_{i=n}^{\infty} (P_0 - P_0^*)^i\right]\left[\sum_{i=0}^{M-1} P^i X\right] + (P_0^n - P_0^*)\left(\sum_{i=0}^{m-1} P^i r + P^m X\right). \blacksquare$$

Now we apply Lemma 5.6 in the above example. First, we note that

$$L(h)X = L(f)L(g)X = r(f) + P(f)r(g) + P(f)P(g)X \qquad (5.19)$$

and $r(h) = r(f) + P(f)r(g)$ and $P(h) = P(f)P(g)$. Applying (a) of Lemma 3.1 we have

$$V^{2n}(\pi, 0) = (2n-1)\begin{bmatrix} \frac{6}{5} \\ \frac{6}{5} \end{bmatrix} + \begin{bmatrix} \frac{58}{35} \\ \frac{6}{7} \end{bmatrix} + \varepsilon(n, h) \qquad (5.20)$$

$$V^{2n+1}(\pi, 0) = L(g)V^{2n}(\pi, 0) = \begin{bmatrix} \frac{3}{5} \\ \frac{3}{5} \end{bmatrix} + V^{2n}(\pi, 0) + \varepsilon'(n, f) \qquad (5.21)$$

and

$$\lim_{n \to \infty} [V^n(\pi, 0) - V^n(g, 0)] = \begin{bmatrix} \frac{2}{35} \\ \frac{2}{35} \end{bmatrix} \qquad (5.22)$$

We shall show below the relationship between the optimal policy and the optimal stationary strategy f^∞.

LEMMA 5.7. *Let f^∞ be the optimal stationary strategy. Then the policy f^∞ is $mu(f) + A(f)$ optimal for some m.*

PROOF.

$$L(f)(mu(f) + A(f)) - L(g)(mu(f) + A(f))$$
$$= m(u(f) - P(g)u(f)) + L(f)A(f) - L(g)A(f) \qquad (5.23)$$

Hence from Lemma 5.5, for $m > m_0$, we have

$$L(g)(mu(f) + A(f)) \leqslant L(f)(mu(f) + A(f))$$

for any $g \in F$. But from Lemma 5.4, we have

$$V^n(g^\infty, mu(f) + A(f)) = (m+n)u(f) + A(f) \qquad (5.24)$$

Thus for any policy $\pi = (\ldots, g_n, \ldots, g_2, g_1)$ and for all n, we have

$$\begin{aligned}
V^n(\pi, mu(f) + A(f)) &= L(g_n) \cdots L(g_1)(mu(f) + A(f)) \\
&\leqslant L^n(f)(mu(f) + A(f)) \\
&= V^n(f^\infty, mu(f) + A(f)). \quad \blacksquare
\end{aligned} \qquad (5.25)$$

THEOREM 5.8. *Let π be the optimal policy and f^∞ be the optimal stationary strategy. Then $V^n(\pi, 0) - V^n(f^\infty_{\mathbb{E}}, 0)$ is bounded uniformly in n.*

PROOF. From Definition 5.2, we have $V^n(\pi, 0) \geqslant V^n(f^\infty, 0)$, which shows that $V^n(\pi, 0) - V^n(f^\infty, 0)$ is bounded below. Suppose it is unbounded above. Then for any X, $V^n(\pi, X) - V^n(f^\infty, X)$ is bounded, since

$$\begin{aligned}
V^n(\pi, X) &- V^n(f^\infty, X) \\
&= V^n(\pi, 0) - V^n(f^\infty, 0) + (P(f_n) \cdots P(f_1) - P^n(f))X
\end{aligned}$$

and the last term is bounded uniformly in n. But setting $X = mu(f) + A(f)$ in Lemma 5.7, we have $V^n(\pi, X) - V^n(f^\infty, X) \leqslant 0$, which contradicts the hypothesis. $\blacksquare$

COROLLARY 5.9. $\lim_{n \to \infty} V^n(\pi, 0)/n = u(f)$ *and $V^n(\pi, 0) - nu(f)$ is bounded uniformly in n.*

PROOF. We note that

$$V^n(\pi, 0) - nu(f) = (V^n(\pi, 0) - V^n(f^\infty, 0)) + (V^n(f^\infty, 0) - nu(f)) \quad (5.26)$$

The two parentheses on the right-hand side are both uniformly bounded from the note of Lemmas 3.6 and 5.6. $\blacksquare$

COROLLARY 5.10. *If g occurs infinity often in π, then $P(g)u(f) = u(f)$, where f^∞ is the optimal strategy.*

PROOF

$$\begin{aligned}
(n + 1)^{-1}V^{n+1}(\pi, 0) &= (n + 1)^{-1}r(f_{n+1}) \\
&+ n/(n + 1)^{-1}P(f_{n+1})[n^{-1}V^n(\pi, 0)]
\end{aligned} \qquad (5.27)$$

Let $n \to \infty$ through a sequence for which $f_{n_i+1} = g$, and use Corollary 5.9. Thus we have $u(f) = P(g)u(f)$. $\blacksquare$

The remainder of this section is devoted to showing that $V^n(\pi, 0)$ is *asymptotically periodic*, that is, for some fixed M,

$$\lim_{n \to \infty} [V^{nM+l}(\pi, 0) - (nM + l)u(f)]$$

exists for all l.

It suffices to prove that $V^n(\pi_X, X)$ is asymptotically periodic where π_X is X-optimal, and X is such that if g occurs in π_X then $P(g)u(f) = u(f)$. This is so because $V^{n+m}(\pi, 0) = V^n(\pi', V^m(\pi, 0))$ where $\pi' = (\ldots, f_{m+2}, f_{m+1})$. Let $X = V^m(\pi, 0)$ and choose m large enough so that those g's that occur only finitely often in π occur only as $f_1, \ldots, f_m$ and not as f_i, $i > m$. The reduction is obtained in Corollary 5.10.

We can then replace each $r(g)$ by $r(g) - u(f)$. In this way we only need to show that $\lim\limits_{n \to \infty} V^{nM+l}(\pi_X, X)$ exists as $u(f) = 0$.

Let Y be a limit point of $\{V^n(\pi_X, X)\}$. (There must be one by Theorem 5.8.) It then suffices to show that for some M, $V^M(\pi_Y, Y) = Y$, where π_Y is Y-optimal.

LEMMA 5.11. *If f_t ($t \in T$) is a set of real-valued functions with*
$$|f_t(a) - f_t(\ell)| \leqslant C$$
for all $t \in T$, then
$$|\sup_t f_t(a) - \sup_t f_t(\ell)| \leqslant C$$

PROOF. Suppose $\sup_t f_t(a) - \sup_t f_t(\ell) = C + \varepsilon$ for some ε. Let t_0 be such that $f_{t_0}(a) > \sup_t f_t(a) - \varepsilon/2$. Then
$$\sup_t f_t(a) - \sup_t f_t(\ell) < f_{t_0}(a) - f_{t_0}(\ell) + \varepsilon/2 < C + \varepsilon/2 \qquad (5.28)$$
which contradicts the hypothesis. Thus we have
$$\sup_t f_t(a) - \sup_t f_t(\ell) \leqslant C \qquad (5.29)$$
Interchanging a and ℓ, we have
$$\sup_t f_t(\ell) - \sup_t f_t(a) \leqslant C. \blacksquare \qquad (5.30)$$

Let $\|X\|$ be the norm, $\|X\| = \max\limits_i |x_i|$, where x_i is the ith element of X.

We shall give the following two lemmas because they are essential in the proof of the next main theorem.

LEMMA 5.12
$$\|V^n(\pi_u, U) - V^n(\pi_v, V)\| \leqslant \|U - V\| \qquad (5.31)$$
where π_u is U-optimal and π_v is V-optimal.

PROOF. For any policy π, we have
$\|V^n(\pi, U) - V^n(\pi, V)\|$
$= \|V^{n-1}(\pi', r(f_1)) + P(f_n) \cdots P(f_1)U - V^{n-1}(\pi', r(f_1)) - P(f_n) \cdots P(f_1)V\|$
$= \|P(f_n) \cdots P(f_1)(U - V)\|$
$\leqslant \|U - V\| \qquad (5.32)$

where $\pi' = (\ldots, f_n, \ldots, f_3, f_2)$. Applying Lemma 5.11 for the above inequality, we have

$$\|\sup_\pi V^n(\pi, U) - \sup_\pi V^n(\pi, V)\| \leqslant \|U - V\|. \quad \blacksquare \qquad (5.33)$$

LEMMA 5.13. *Let Y be a limit point of $V^n(\pi_X, X)$, where π_X is X-optimal. Then, for some N, $V^N(\pi_X, Y) = Y$.*

PROOF.   Since Y is a limit point of $V^n(\pi_X, X)$, there exists a partial sequence $\{n_i\}$ such that $V^{n_i}(\pi, X) \to Y$ as $i \to \infty$. Hence for sufficiently large i, we have $\|V^n(T^{n_i}\pi_X, V^{n_i}(\pi_X, X)) - V^n(\pi_Y, Y)\| < \varepsilon$, where $T^n\pi$ means the policy $(\ldots, f_{n+2}, f_{n+1})$ for any policy $\pi = (\ldots, f_2, f_1)$. Since n is arbitrary, the set of the limit points of $\{V^n(\pi_Y, Y)\}$ is equal to that of $\{V^n(\pi_X, X)\}$. Thus Y must be a limit point of $V^n(\pi_Y, Y)$. Consequently, either $V^N(\pi_Y, Y) = Y$ for some N or else $V^m(\pi_Y, Y) \neq V^n(\pi_Y, Y)$ for any $m \neq n$. Suppose that the latter holds.

Note that $V^n(\pi_A, A) = A$ from Lemma 5.7 and Lemma 5.4. (Note also that $u(f) = 0$.) Let Y_0 be a limit point of $\{V^n(\pi_Y, Y)\}$ with $\|Y_0 - A\|$ a minimum (which must be attained because the set of limit points of $\{V^n(\pi_Y, Y)\}$ is a compact set).

From Lemma 5.12 we have

$$\|V^n(\pi_{Y_0}, Y_0) - V^n(\pi_A, A)\| = \|V^n(\pi_{Y_0}, Y_0) - A\| \leqslant \|Y_0 - A\|$$

It follows that each limit point of $V^n(\pi_{Y_0}, Y_0)$ is the same distance from A. But since Y is a limit point of $V^n(\pi_Y, Y)$, each point of $V^n(\pi_Y, Y)$ is a limit point of $V^n(\pi_Y, Y)$. Hence all are the same distance from A.

Note that $A + \Xi$, where each element of Ξ is ξ, has the property that $V^n(\pi_{A+\Xi}, A + \Xi) = A + \Xi$. Hence, by the same argument, all $V^n(\pi_Y, Y)$ are the same distance from $A + \Xi$.

Let an infinite number of $V^n(\pi_Y, Y)$ be such that

$$|[V^n(\pi_Y, Y) - A]_{i_1}| = \|V^n(\pi_Y, Y) - A\| \qquad (5.34)$$

for some i_1, where $[X]_i$ denotes the ith element of X. Then there is a ξ_1 such that an infinite number of these have the same distance from $A - \Xi_1$, where each element of Ξ_1 is ξ_1. If this distance is zero, the proof is finished. If not, there is an $i_2 \neq i_1$ with

$$|[V^n(\pi_Y, Y) - (A - \Xi_1)]_{i_2}| = \|V^n(\pi_Y, Y) - (A - \Xi)\|$$

for infinitely many of the $V^n(\pi_Y, Y)$ selected. Thus, infinitely many of these have $[V^n(\pi_Y, Y)]_{i_2} = \xi_2$ and $[V^n(\pi_Y, Y)]_{i_1} = \xi_1$. Continuing in this manner, after at most N steps, we have infinitely many $V^n(\pi_Y, Y)$ agreeing in all elements. But this contradicts the assumption that $V^m(\pi_Y, Y) \neq V^n(\pi_Y, Y)$ for $m \neq n$. ∎

We come to the main theorem that $V^n(\pi, 0)$ is *asymptotically periodic*.

THEOREM 5.14.   *If π is an optimal policy, then there is an M such that*

$$\lim_{n \to \infty} [V^{nM+l}(\pi, 0) - (nM + l)u(f^*)] \tag{5.35}$$

exists for any l, where $(f^*)^\infty$ is an optimal strategy.

PROOF.   For an optimal policy $\pi = (\ldots, f_2, f_1)$, we suppose that f occurs only finitely many times. Then we can choose m such that f occurs only in $f_1, \ldots, f_m$. We have

$$V^{n+m}(\pi, 0) = V^n(T^m\pi, V^m(\pi, 0)) \tag{5.36}$$

where $T^m\pi = (\ldots, f_{m+2}, f_{m+1})$. Now we set $X = V^m(\pi, 0)$ and denote X-optimal $T^m\pi$ by π_X. We can now show that $V^n(\pi_X, X)$ is asymptotically periodic.

To show that $V^n(\pi_X, X)$ is asymptotically periodic, we show that there is an M such that $V^M(\pi_Y, Y) = Y$, where Y is a limit point of $V^n(\pi_X, X)$, and π_Y is Y-optimal. Because Y is a limit point of $V^n(\pi_X, X)$, there exists a partial sequence $\{n_i\}$ such that $\|V^{n_i}(\pi_X, X) - Y\| < \varepsilon$ for any $\varepsilon > 0$. Applying Lemma 5.12, we have

$$\|V^{n_i+nM}(\pi_X, X) - Y\| = \|V^{nM}(T^{n_i}\pi_X, V^{n_i}(\pi_X, X)) - V^{nM}(\pi_Y, Y)\|$$

$$\leqslant \|V^{n_i}(\pi_X, X) - Y\| < \varepsilon \tag{5.37}$$

From Lemma 5.13, we have $V^M(\pi_Y, Y) = Y$ for some M. ∎

COROLLARY 5.15.   *Given $\varepsilon > 0$, there is a policy $\pi' = (\ldots, f_n, \ldots, f_1)$ with $f_{m+N} = f_m$ for $m > m_0$ such that*

$$\|V^n(\pi, 0) - V^n(\pi', 0)\| < \varepsilon \tag{5.38}$$

for optimal π, uniformly in n.

PROOF.   Let Y be a limit point of $V^n(\pi, 0)$. For sufficiently large m_0, we have $\|V^{m_0}(\pi, 0) - Y\| < \varepsilon$ for $\varepsilon > 0$. Setting $X = V^{m_0}(\pi, 0)$ and $T^{m_0}\pi = \pi_X$, we have

$$\|V^n(\pi_X, X) - V^n(\pi_Y, Y)\| < \varepsilon \tag{5.39}$$

where π_Y is a periodic Y-optimal policy. Thus we obtain $\pi' = (\pi_Y, f_{m_0}, \ldots, f_1)$.

∎

References and Comments

The results of this chapter have all been given by Brown [19]. Brown [19] also has studied a special case where each element of $P(f)$ is positive, that is, $P(f)$ is regular. For the special case, he has given the strengthened theorem that $V^n(\pi, 0) - nu(f) - A(f)$ converges as $n \to \infty$ to a limit vector with identical elements.

For the theory of dynamic programming, see Bellman [7, 8] and Bellman and Dreyfus [9].

Derman and Klein [43] have formulated the Markovian decision process over a finite horizon as a linear programming problem, but we omitted this discussion here.

For the process discussed in this chapter, the turnpike theorem is also interesting (see Shapiro [104]).

Chapter 6

Semi-Markovian Decision Processes

6.1. Introduction

In the preceding chapters we discussed decision processes in which the decisions are made synchronously in each time period. In this chapter we discuss decision processes in which the sojourn time in each state is random and the decisions are made asynchronously. That is to say, our preceding models are based on a discrete-time Markov chain. We can, however, consider a decision process based on a continuous-time stochastic process.

An extension of a discrete-time Markov chain is a continuous-time Markov process. A decision process based on a continuous-time Markov process has been studied by Howard [63]. In this chapter, however, we treat a decision process based on a semi-Markov process, which includes a continuous-time Markov process as a special case.

A *semi-Markov process*, or a *Markov renewal process*, is a marriage of Markov processes and renewal processes. A semi-Markov process, roughly speaking, is a stochastic process that moves from one state to another with given probability laws; but the sojourn time in a state is a random variable with distribution depending on that state and the next state visited. Semi-Markov processes include a renewal process, a discrete-time Markov chain, and a continuous-time Markov process as special cases. That is, a renewal process is a semi-Markov process with one state. A discrete-time Markov chain is a semi-Markov process in which every sojourn-time distribution is degenerate at unit time, and a continuous-time Markov process is a semi-Markov process in which all distributions are exponential.

In this chapter we shall treat a decision process based on a semi-Markov process or a Markov renewal process—that is, a *semi-Markovian decision process*, or a *Markov renewal program*.

In Section 6.2 we shall study the properties of semi-Markov processes, and in Section 6.3, we shall further consider semi-Markov processes with returns. In the successive sections we shall introduce semi-Markovian decision processes and show that semi-Markovian decision processes may be formulated as linear programming problems for discounted and nondiscounted models. In this discussion we shall derive the corresponding policy iteration algorithms from these linear programming problems.

6.2. Semi-Markov Processes

In this section we shall present information about semi-Markov processes that will be of use later. Semi-Markov processes were first discussed by Lévy, Smith, and Takács, in 1954.

Consider a system that moves from state to state. Let us define a stochastic process as $\{Z_t; t \geqslant 0\}$, where $Z_t = i$ denotes that the system is in state i at time t. The states are denoted by integers $i = 1, 2, \ldots, N \in S$, and in general the state space S is countable; however, in this chapter we shall consider only the case where S is *finite*, although some of the results are valid for the countable case. Note that later we shall introduce the concept of regularity for each state.

The probability laws of state transitions obey an $N \times N$ Markov matrix $P = [p_{ij}]$, which is called an "imbedded Markov chain," where p_{ij} is the probability from state i to state j. Further, the sojourn time in any state i is a random variable, with distribution $F_{ij}(t)$ that depends both on that state and on the next state j visited. We define

$$Q_{ij}(t) = p_{ij}F_{ij}(t) \qquad \text{for} \quad i, j \in S \tag{6.1}$$

where $Q_{ij}(t)$ satisfies

(a) $$Q_{ij}(0) = 0 \qquad \text{for} \quad i, j \in S \tag{6.2}$$

(b) $$\sum_{j \in S} Q_{ij}(\infty) = \sum_{j \in S} p_{ij} = 1 \qquad \text{for} \quad i \in S \tag{6.3}$$

We define an $N \times N$ matrix Q with element $Q_{ij}(t)$, which is called a "matrix of transition distributions." Further, we define

$$H_i(t) = \sum_{j \in S} Q_{ij}(t) \qquad \text{for} \quad i \in S \tag{6.4}$$

which is a distribution of the sojourn time in state i disregarding the next state. Hence, it is called a "distribution of unconditional sojourn time," or an "*unconditional* distribution." Let a be an initial distribution ($1 \times N$ row vector);

$$a = (a_1, a_2, \ldots, a_N) \tag{6.5}$$

where

$$\sum_{i \in S} a_i = 1, \qquad a_i \geqslant 0 \qquad \text{for} \quad i \in S \tag{6.6}$$

Then the stochastic process $\{Z_t; t \geqslant 0\}$ is called a semi-Markov process determined by

$$(N, a, Q) \tag{6.7}$$

When we are concerned with an N-dimensional renewal quantity

$$\mathfrak{N}(t) = (\mathfrak{N}_1(t), \mathfrak{N}_2(t), \ldots, \mathfrak{N}_N(t)) \tag{6.8}$$

the stochastic process $\{\mathfrak{N}(t); t \geqslant 0\}$ is called a *Markov renewal process*.

A semi-Markov process is *regular* if each state is entered only a finite number of times within a finite time with probability 1, that is $P_r[\mathfrak{N}_i(t) < \infty] = 1$ for each $i \in S$ and $t \geqslant 0$. Thus a regular semi-Markov process always has a finite number of transitions within a finite time. In this chapter we consider only a regular semi-Markov process throughout this chapter. We suppose that each state is defined in such a way that the semi-Markov process is regular.

We define

$$P_{ij}(t) = P_r[Z_t = j | Z_0 = i] \qquad \text{for all} \quad i, j \in S \tag{6.9}$$

which denotes the probability that the system is in state j at time t given that the system is in state i at time zero. Further we define

$$G_{ij}(t) = P_r[\mathfrak{N}_j(t) > 0 | Z_0 = i] \qquad \text{for all} \quad i, j \in S \tag{6.10}$$

$$M_{ij}(t) = E[\mathfrak{N}_j(t) | Z_0 = i] \qquad \text{for all} \quad i, j \in S \tag{6.11}$$

where $E[\ ]$ denotes the expectation. Here $G_{ij}(t)$ is the first passage time distribution from state i to state j, and $M_{ij}(t)$ is the mean number of visits to state j up to time t starting in state i, which refers to a renewal function in renewal theory (see, for example, Cox [24] and Smith [108]).

From renewal-theoretic considerations we immediately have

$$P_{ij}(t) = (1 - H_i(t))\delta_{ij} + \sum_{k \in S} Q_{ik}(t) * P_{kj}(t)$$

$$= (1 - H_i(t))\delta_{ij} + G_{ij}(t) * P_{jj}(t) \tag{6.12}$$

$$G_{ij}(t) = Q_{ij}(t) + \sum_{\substack{k \in S \\ k \neq j}} Q_{ik}(t) * G_{kj}(t) \tag{6.13}$$

$$M_{ij}(t) = G_{ij}(t) + G_{ij}(t) * M_{jj}(t)$$

$$= Q_{ij}(t) + \sum_{k \in S} Q_{ik}(t) * M_{kj}(t) \tag{6.14}$$

where $*$ denotes the convolution. To solve the above equations with respect to $P_{ij}(t)$, $G_{ij}(t)$ and $M_{ij}(t)$, we apply the Laplace–Stieltjes transforms. Let the small letter with argument s denotes the Laplace–Stieltjes transform of the corresponding quantity, for example, $p_{ij}(s)$ is the Laplace–Stieltjes transform of $P_{ij}(t)$. Further, let $q(s)$ and $m(s)$ be $N \times N$ matrices with elements $q_{ij}(s)$ and $m_{ij}(s)$, respectively.

Taking the Laplace–Stieltjes transforms for (6.12), (6.13), and (6.14), and solving for each quantity, we have

$$m(s) = [I - q(s)]^{-1}q(s) = [I - q(s)]^{-1} - I \tag{6.15}$$

$$g_{ij}(s) = m_{ij}(s)/[1 + m_{jj}(s)] \tag{6.16}$$

$$p_{ij}(s) = p_{jj}(s)g_{ij}(s) \qquad \text{for} \quad i \neq j \tag{6.17}$$

$$p_{jj}(s) = [1 - h_j(s)]/[1 - g_{jj}(s)] \tag{6.18}$$

Thus, successively inverting the above equations, we have the required quantities, $M_{ij}(t)$, $G_{ij}(t)$, and $P_{ij}(t)$.

We can now give the state classification for a semi-Markov process. Two states i and j are said to *communicate* if $G_{ij}(\infty)G_{ji}(\infty) > 0$. Since the communication relation is an equivalence, we may apply the usual state classification for finite Markov chains, as in Kemeny–Snell [70].

Let

$$\ell_{ij} = \int_0^\infty t \, dF_{ij}(t) \qquad \text{for} \quad i, j \in S \tag{6.19}$$

and

$$\eta_i = \int_0^\infty t \, dH_i(t) = \sum_{j \in S} p_{ij} \int_0^\infty t \, dF_{ij}(t) = \sum_{j \in S} p_{ij}\ell_{ij} \qquad \text{for} \quad i \in S \tag{6.20}$$

be mean values of $F_{ij}(t)$ and $H_i(t)$, respectively. All these mean values are assumed to be finite. Further we shall give the additional assumptions if necessary.

Let μ_{ij} and $\mu_{ij}^{(2)}$ be the first and second moments of $G_{ij}(t)$. From mean-theoretic considerations, we have

$$\mu_{ij} = \sum_{\substack{k \in S \\ k \neq j}} p_{ik}\mu_{kj} + \eta_i \tag{6.21}$$

$$\mu_{ij}^{(2)} = \sum_{\substack{k \in S \\ k \neq j}} p_{ik}[\mu_{kj}^{(2)} + 2\ell_{ik}\mu_{kj}] + \eta_i^{(2)} \tag{6.22}$$

where

$$\eta_i^{(2)} = \int_0^\infty t^2 \, dH_i(t) \qquad \text{for} \quad i \in S \tag{6.23}$$

If the imbedded Markov chain is *ergodic* and we let π_j be the limiting probability of the imbedded Markov chain, we have

$$\mu_{jj} = \sum_{k \in S} \pi_k \eta_k / \pi_j \tag{6.24}$$

$$\mu_{jj}^{(2)} = \Big[\sum_{k \in S} \pi_k \eta_k^{(2)} + 2 \sum_{\substack{k \in S \\ k \neq j}} \pi_i \ell_{ik}\mu_{kj}\Big]/\pi_j \tag{6.25}$$

where μ_{jj} and $\mu_{jj}^{(2)}$ are the first and second moments of the recurrence time in state j.

We consider the limiting (or steady state) probability $P_{ij}(\infty)$. In general, we have

$$P_{ij}(\infty) = G_{ij}(\infty)\eta_j / \mu_{jj} \tag{6.26}$$

In particular, if the semi-Markov process is *ergodic*, from $G_{ij}(\infty) = 1$ for all $i, j \in S$ and (6.24), we have

$$P_{ij}(\infty) = \eta_j / \mu_{jj} = \pi_j \eta_j / \sum_{k \in S} \pi_k \eta_k \tag{6.27}$$

which is independent of any initial distribution. In general we can obtain $G_{ij}(\infty)$ and μ_{jj} for any $i, j \in S$, and thus we can obtain the limiting probability $P_{ij}(\infty)$ for any $i, j \in S$. We omit the results here.

Let n_i be the mean time to absorption, starting in any transient state $i \in T$. Here the word "absorption" means the entrance of some ergodic state. We have

$$n_i = \sum_{j \notin T} p_{ij} \ell_{ij} + \sum_{j \in T} p_{ij}(\ell_{ij} + n_j)$$
$$= \eta_i + \sum_{j \in T} p_{ij} n_j \qquad (6.28)$$

Let n and η be column vectors with n_i and η_i, respectively, where $i \in T$. Then we have

$$n = \eta + P_T n \qquad (6.29)$$

or

$$n = [I - P_T]^{-1} \eta, \qquad (6.30)$$

where P_T is a submatrix of P eliminating all ergodic states.

Finally we shall obtain the asymptotic behavior about the generalized renewal quantities $M_{ij}(t)$. From (6.16), we have

$$m_{ij}(s) = g_{ij}(s)[1 + m_{jj}(s)]$$
$$= g_{ij}(s)\{1 + g_{jj}(s)/[1 - g_{jj}(s)]\} \qquad (6.31)$$

If the states i and j are in the same ergodic set, expanding $g_{ij}(s)$ for small s we have

$$m_{ij}(s) = (s\mu_{jj})^{-1} + \mu_{jj}^{(2)}/2(\mu_{jj})^2 - \mu_{ij}/\mu_{jj} + 0(1) \qquad (6.32)$$

Thus from a Tauberian theorem (Widder [119, p. 192]) we have

$$M_{ij}(t) - t/\mu_{jj} \to \mu_{jj}^{(2)}/2(\mu_{jj})^2 - \mu_{ij}/\mu_{jj} \qquad (6.33)$$

which is a result analogous to an extension of the so-called elementary renewal theorem (see, for example, Cox [24] and Smith [108]). We note that, if the semi-Markov process considered is periodic, the limit of (6.26), (6.27), (6.30), and (6.31) is the sense of Césaro limit; conversely, if the semi-Markov process is nonperiodic, the limit is the usual one.

6.3. Semi-Markov Processes with Returns

In this section we shall consider the returns associated with semi-Markov processes. In the discrete time case we have considered the return r_i per time period, but in the continuous time case we can consider the returns r_i ($i \in S$). That is, when the system is in state i during a unit time, we receive the return r_i. Thus when the system is in state i during a time duration t, we receive the return $r_i t$.

First, we shall consider the discounted case. Since the process is continuous in time, we use a *discount factor* α ($\alpha > 0$) of exponential type. That is, if we have a unit return at any time, we have the return $e^{-\alpha t}$ after a time duration t from that time. Then, if we have the return r_i, the accumulated return between 0 and t is

$$\int_0^t r_i e^{-\alpha \tau} d\tau = \frac{r_i}{\alpha}[1 - e^{-\alpha t}] \tag{6.34}$$

Let $v_i(t)$ be the discounted total expected return up to time t starting in state i at time zero. Then we have similarly the following equation of renewal type with respect to $v_i(t)$:

$$v_i(t) = [1 - H_i(t)]\frac{r_i}{\alpha}[1 - e^{-\alpha t}]$$

$$+ \sum_{j \in S} \int_0^t \left\{ \frac{r_i}{\alpha}[1 - e^{-\alpha \tau}] + e^{-\alpha \tau} v_j(t - \tau) \right\} dQ_{ij}(\tau) \quad \text{for } i \in S \tag{6.35}$$

[Cf. for example, (6.9).] Since the discounted total expected return $v_i(t)$ converges as $t \to \infty$ to a finite value, we set

$$v_i(\alpha) = \lim_{t \to \infty} v_i(t) \qquad \text{for } i \in S \tag{6.36}$$

[note that $v_i(\alpha)$ is a function of α] and, letting $t \to \infty$ in (6.35), we have

$$v_i(\alpha) = (r_i/\alpha)[1 - h_i(\alpha)] + \sum_{j \in S} q_{ij}(s)v_j(\alpha) \qquad \text{for } i \in S \tag{6.37}$$

where $h_i(\alpha)$ and $q_{ij}(\alpha)$ are the Laplace–Stieltjes transforms of $H_i(t)$ and $Q_{ij}(t)$ evaluated at $s = \alpha$, respectively. Setting

$$\rho_i(\alpha) = (r_i/\alpha)[1 - h_i(\alpha)] \qquad \text{for } i \in S \tag{6.38}$$

and defining two $N \times 1$ column vectors

$$\rho(\alpha) = \begin{bmatrix} \rho_1(\alpha) \\ \cdot \\ \cdot \\ \cdot \\ \rho_N(\alpha) \end{bmatrix}, \qquad v(\alpha) = \begin{bmatrix} v_1(\alpha) \\ \cdot \\ \cdot \\ \cdot \\ v_N(\alpha) \end{bmatrix} \tag{6.39}$$

we have

$$v(\alpha) = \rho(\alpha) + q(\alpha)v(\alpha) \tag{6.40}$$

or

$$v(\alpha) = [I - q(\alpha)]^{-1}\rho(\alpha), \tag{6.41}$$

since we know that $I - q(\alpha)$ is nonsingular for $\alpha > 0$. Thus we have the discounted total expected return starting in each state $i \in S$. If the system starts in an initial distribution a in (6.5), we have

$$av(\alpha) = a[I - q(\alpha)]^{-1}\rho(\alpha) \tag{6.42}$$

Secondly, we shall consider the nondiscounted case. We have discussed two approaches of the nondiscounted model in the discrete time case. For the continuous model we also have two approaches; the first treats it as a limiting case of $\alpha > 0$; the second deals directly in terms of the long-run average per unit time. In the first approach, we note that from (6.41) and (6.15), we have

$$v(\alpha) = [I + m(\alpha)]\rho(\alpha) \qquad (6.43)$$

We now assume that the semi-Markov process is *ergodic*. Using (6.32), the asymptotic behaviour $m(\alpha)$ for a small α, and noting that $\lim_{\alpha \to 0} \rho_i(\alpha) = r_i \eta_i$,

we have

$$v_i(\alpha) = \frac{g}{\alpha} + v_i + 0(1) \qquad \text{for} \quad i \in S \qquad (6.44)$$

where

$$g = \sum_{j \in S} \frac{r_j \eta_j}{\mu_{jj}} = \sum_{j \in S} \pi_j \eta_j r_j / \sum_{j \in S} \pi_j \eta_j \qquad (6.45)$$

and

$$v_i = r_i \eta_i + \sum_{j \in S} r_j \eta_j \left[\frac{\mu_{jj}^{(2)}}{2(\mu_{jj})^2} - \frac{\mu_{ij}}{\mu_{jj}} + \delta_{ij} \right] \qquad (6.46)$$

Here we assume that $\mu_{jj}^{(2)}$ is finite. Note that g is the long-run *average return per unit time* and is the result of the average of r_i $(i \in S)$ with respect to the limiting probabilities in (6.27).

It is easy to extend the result of (6.44) to any ergodic state. For any ergodic state i belonging to an ergodic set $E_v \subset S$, we have a result similar to (6.44) except that the summation is taken over E_v instead of S. For (6.45) and (6.46) the summation is also taken over E_v.

Applying the same approach, we have for each state (see Fox [59])

$$v(\alpha) = u/\alpha + v + 0(1) \qquad (6.47)$$

where u and v are $N \times 1$ column vectors. This corresponds to (4.2) in the discrete-time case.

We next consider the second approach, that of treating directly the long-run average per unit time. Let $v_i(t)$ be the total expected return up to time t starting in state i at time zero. Then $v_i(t)$ satisfies the following equation of renewal type derived in the same way as we have derived the equation for $P_{ij}(t)$. That is, we have

$$v_i(t) = [1 - H_i(t)]r_i t + \sum_{j \in S} \int_0^t [r_i \tau + v_j(t - \tau)] \, dQ_{ij}(\tau) \quad \text{for} \quad i \in S \qquad (6.48)$$

We shall first assume that a semi-Markov process is *ergodic*. In this case we shall consider only the average return per unit time in the steady state.

Therefore, for a sufficiently large t, the first term of the right-hand side of (6.48) tends to zero because of the finiteness of the mean of $H_i(t)$ and because $[1 - H_i(t)]$ converges to zero more rapidly than $r_i t$ diverges. Consequently, for a sufficiently large t, we have

$$v_i(t) = \sum_{j \in S} \int_0^t \left[r_i \tau + v_j(t - \tau) \right] dQ_{ij}(\tau)$$

$$= \sum_{j \in S} \int_0^t r_i \tau \, dQ_{ij}(\tau) + \sum_{j \in S} \int_0^t v_j(t - \tau) \, dQ_{ij}(\tau) \qquad \text{for} \quad i \in S \qquad (6.49)$$

The first term of the right-hand side of (6.49) tends to $r_i \eta_i$ as $t \to \infty$, then we have for a sufficiently large t,

$$v_i(t) = r_i \eta_i + \sum_{j \in S} \int_0^t v_j(t - \tau) \, dQ_{ij}(\tau) \qquad \text{for} \quad i \in S \qquad (6.50)$$

Let $v_i(s)$ be the Laplace transform of $v_i(t)$. We define two $N \times 1$ column vectors

$$v(s) = \begin{bmatrix} v_1(s) \\ \cdot \\ \cdot \\ \cdot \\ v_N(s) \end{bmatrix}, \qquad r = \begin{bmatrix} r_1 \eta_1 \\ \cdot \\ \cdot \\ \cdot \\ r_N \eta_N \end{bmatrix} \qquad (6.51)$$

Applying the Laplace transforms to (6.50), we have

$$v(s) = (1/s)r + q(s)v(s) \qquad (6.52)$$

or

$$v(s) = (1/s)[I - q(s)]^{-1} r \qquad (6.53)$$

Using (6.15), we have

$$v(s) = (1/s)[I + m(s)]r \qquad (6.54)$$

The total expected return up to time t starting in an initial distribution a is $av(t)$, where $v(t)$ is the $N \times 1$ column vector with ith element $v_i(t)$. Since $\lim_{s \to 0} s^2 a[I + m(s)]r = \sum_{j \in S} (r_j \eta_j / \mu_{jj})$, from a Tauberian theorem (Widder [119, p. 192]) we have

$$\lim_{t \to \infty} \frac{av(t)}{t} = \sum_{j \in S} \frac{r_j \eta_j}{\mu_{jj}} = \sum_{j \in S} \pi_j r_j \eta_j / \sum_{j \in S} \pi_j \eta_j \qquad (6.55)$$

which is independent of an initial distribution a. Thus $\mathfrak{g}$ is the long-run average per unit time which has been obtained in (6.45).

We shall next consider the *terminating process* (see Section 3.7). Since we consider only the total expected return before absorption, it is sufficient for this case that the semi-Markov process considered is absorbing. Hence we assume that state 1 is absorbing and other states are transient. Let $v_i(t)$ be the total expected return up to time t. Then we obtain the following equation of renewal type in the same way as we have derived (6.48):

$$v_i(t) = [1 - H_i(t)]r_i t + \int_0^t r_i \tau \, dQ_{i1}(\tau) + \sum_{j=2}^N \int_0^t [r_i \tau + v_j(t - \tau)] \, dQ_{ij}(\tau)$$
$$\text{for} \quad i = 2, \ldots, N \qquad (6.56)$$

Furthermore

$$v_i = \lim_{t \to \infty} v_i(t) \qquad \text{for} \quad i = 2, \ldots, N \qquad (6.57)$$

which is valid since the system reaches the absorbing state with probability 1. Assuming $t \to \infty$ in (6.56), we have

$$v_i = \sum_{j \in S} r_i \int_0^\infty \tau \, dQ_{ij}(\tau) + \sum_{j=2}^N v_j \int_0^\infty dQ_{ij}(\tau)$$
$$= r_i \eta_i + \sum_{j=2}^N v_j p_{ij} \qquad \text{for} \quad i = 2, \ldots, N \qquad (6.58)$$

Introducing two $(N - 1) \times 1$ column vectors

$$v' = \begin{bmatrix} v_2 \\ \cdot \\ \cdot \\ \cdot \\ v_N \end{bmatrix}, \qquad r' = \begin{bmatrix} r_2 \eta_2 \\ \cdot \\ \cdot \\ \cdot \\ r_N \eta_N \end{bmatrix} \qquad (6.59)$$

we can rewrite (6.58) in matrix form as follows:

$$v' = r' + P_1 v'$$

where P_1 is the $(N - 1) \times (N - 1)$ submatrix of P eliminating the first row and column. Thus we have

$$v' = [I - P_1]^{-1} r' \qquad (6.60)$$

where $[I - P_1]^{-1}$ is the *fundamental matrix* of the absorbing Markov chain in the discrete-time case (see Kemeny and Snell [70, p. 46]). Therefore the total expected return before absorption, starting in an initial distribution a, is

$$a'v' = a'[I - P_1]^{-1} r' \qquad (6.61)$$

where $a' = (a_2, a_3, \ldots, a_N)$.

We have defined the return structure to be the return r_i when the system is in state i during a unit time. Now we shall consider the generalized returns (see Jewell [65]). When the system is in state i, and the next visiting state j, a return is accumulated according to the arbitrary function $R_{ij}(t|\tau)$, depending on i and j, on the sojourn time $\tau(i, j)$, and on the clock time t since the beginning of the transition interval $[0 \leqslant t \leqslant \tau(i, j)]$. We shall assume that $R_{ij}(0|\tau) = 0$, and will denote the total return at the end of the interval by $R_{ij}(\tau|\tau) = R_{ij}(\tau)$. Returns from successive transitions are additive.

For the discounted model, the average, one-step-discounted return starting in state i is

$$\rho_i(\alpha) = \sum_{j \in S} \int_0^\infty dQ_{ij}(\tau) \int_0^\tau e^{-\alpha x} dx\, R_{ij}(x|\tau) \qquad \text{for} \quad i \in S \qquad (6.62)$$

which corresponds to (6.38). The return structure previously considered above is a special case $R_{ij}(x|\tau) = r_i x$. Thus the return structures are essentially the same ones for the discounted model.

For the undiscounted model we have in general $\rho_i(\alpha) \to \rho_i$ (constant) as $\alpha \to 0$, which corresponds to $\rho_i = r_i \eta_i$, but the bias term (6.46) will change slightly (see Jewell [66]).

Consequently, in both cases it suffices to consider only the simple return structures r_i ($i \in S$) instead of $R_{ij}(t|\tau)$.

6.4. Semi-Markovian Decision Processes with Discounting

We shall consider the semi-Markovian decision processes with discounting using the results of the preceding section.

The semi-Markovian decision processes with which we are now concerned may be described as follows: When the system is in state i, we have K_i actions in each state $i \in S$. If we choose an action k in state $i \in S$, the system obeys the probability law $Q_{ij}^k(t) = p_{ij}^k F_{ij}^k(t)$ ($j \in S$), where p_{ij}^k is the transition probability from state i to state j, and $F_{ij}^k(t)$ is the sojourn time in state i knowing that the next visiting state is j. We get the return r_i^k when the system is in state i during a unit time. In other words, we have K_i selections in each state i, that is, we have $k = 1, \ldots, K_i$ actions

$$Q_{ij}^k(t), \qquad (p_{ij}^k, F_{ij}^k(t)), \qquad \text{and} \quad r_i^k \qquad (j \in S) \qquad (6.63)$$

for each $i \in S$.

Then, what strategy—that is, what sequence of actions in each transition epoch for each state over all $t \geqslant 0$—maximizes the total expected return (or the average return per unit time) starting in an initial distribution?

As we have seen in the preceding section, the total expected return is finite if the process is discounted or terminating, and is usually infinite if the process is nondiscounted. Thus we shall consider the total expected return if it is finite and the average return per unit time if it is infinite. In this section we shall first treat the discounted process.

Note that we can consider the semi-Markovian decision process over a finite time horizon. But for this model, the sojourn time in any state is a random variable and we cannot ascertain the transition epoch. Thus we can expect no results, so far as we know, except those based on an approximation approach.

For our model with an infinite time horizon we define a *stationary strategy* such that for each state we take the identical action independent of the history of previous states, transition times, and decisions. In general we should consider a nonstationary strategy depending on the history of states and transition times, and decisions.

Let $v(\alpha, \pi)$ be the discounted total expected return ($N \times 1$ column vector) starting in each state i using any policy π, where a discount factor $\alpha > 0$.

DEFINITION 6.1. *A strategy* π^* *is called* α-optimal *if* $v(\alpha, \pi^*) \geqslant v(\alpha, \pi)$ *for all* π, *where* α ($\alpha > 0$) *is fixed.*

THEOREM 6.2. *There is an* α-optimal *strategy which is stationary.*

The proof of this theorem has been established via a contraction mapping (Denardo [31]) and will be stated in Example 4 of Section 8.5.

From Theorem 6.2 we may find α-optimal strategies within stationary strategies whose number is finite. Then we can use immediately the preceding result (6.41).

We now develop a linear programming formulation for the semi-Markovian decision process with discounting. From (6.41) we have for any stationary strategy π,

$$v(\alpha, \pi) = [I - q(\alpha)]^{-1}\rho(\alpha) \qquad (6.64)$$

The discounted total expected return starting in an initial distribution a is

$$av(\alpha, \pi) = a[I - q(\alpha)]^{-1}\rho(\alpha)$$
$$= \sum_{i \in S} \sum_{j \in S} a_i \mu_{ij}(\alpha)\rho_j(\alpha) \qquad (6.65)$$

where

$$[I - q(\alpha)]^{-1} = [\mu_{ij}(\alpha)] \qquad (6.66)$$

and

$$[I - q(\alpha)]^{-1}[I - q(\alpha)] = I$$

or

$$\sum_{j \in S} \mu_{ij}(\alpha)(\delta_{jl} - q_{jl}(\alpha)) = \delta_{il} \qquad (6.67)$$

It is convenient to extend the range of decisions to include randomized (or mixed) strategies. Let the probability that we make an action k when the system is in state i be $d_i^k (i \in S, k \in K_i)$, which is independent of time t since we consider only the stationary strategies. It is clear that

$$d_i^k \geqslant 0, \qquad \sum_{k \in K_i} d_i^k = 1 \qquad \text{for} \quad i \in S, \quad k \in K_i \qquad (6.68)$$

7

Using d_i^k, we have for any stationary strategy π,

$$av(\alpha, \pi) = \sum_{i \in S} \sum_{j \in S} \sum_{k \in K_j} a_i \mu_{ij}(\alpha) \rho_j^k(\alpha) d_j^k \qquad (6.69)$$

where

$$\rho_j^k(\alpha) = (r_i^k/\alpha)[1 - h_i^k(\alpha)] \qquad (6.70)$$

$$h_i^k(\alpha) = \sum_{j \in S} \int_0^\infty e^{-\alpha t} \, dQ_{ij}^k(t) = \sum_{j \in S} q_{ij}^k(\alpha) \qquad (6.71)$$

Note that $\mu_{ij}(\alpha)$ depends on d_i^k since $[I - q(\alpha)]$ is determined by d_i^k. Now we define

$$x_j^k = \sum_{i \in S} a_i \mu_{ij}(\alpha) \, d_j^k \geqslant 0 \qquad (6.72)$$

Rewriting (6.69) by using x_j^k, we have

$$av(\alpha, \pi) = \sum_{j \in S} \sum_{k \in K_j} \rho_j^k(\alpha) x_j^k \qquad (6.73)$$

We calculate

$$\begin{aligned}
\sum_{j \in S} \sum_{k \in K_j} (\delta_{jl} - q_{jl}^k(\alpha)) x_j^k &= \sum_{i \in S} \sum_{j \in S} \sum_{k \in K_j} (\delta_{jl} - q_{jl}^k(\alpha)) a_i \mu_{ij}(\alpha) \, d_j^k \\
&= \sum_{i \in S} \sum_{j \in S} a_i \mu_{ij}(\alpha) (\sum_{k \in K_j} (\delta_{jl} - q_{jl}^k(\alpha)) \, d_j^k) \\
&= \sum_{i \in S} a_i [\sum_{j \in S} \mu_{ij}(\alpha)(\delta_{jl} - q_{jl}(\alpha))] \\
&= \sum_{i \in S} a_i \delta_{il} \qquad \text{[from (6.67)]} \\
&= a_l \qquad \text{for} \quad l \in S \qquad (6.74)
\end{aligned}$$

that is,

$$\sum_{k \in K_j} x_j^k - \sum_{i \in S} \sum_{k \in K_i} q_{ij}^k(\alpha) x_i^k = a_j \qquad \text{for} \quad j \in S \qquad (6.75)$$

Thus we have the following *linear programming problem:*

$$\max \sum_{j \in S} \sum_{k \in K_j} \rho_j^k(\alpha) x_j^k \qquad (6.76)$$

subject to

$$\sum_{k \in K_j} x_j^k - \sum_{i \in S} \sum_{k \in K_i} q_{ij}^k(\alpha) x_i^k = a_j \qquad \text{for} \quad j \in S \qquad (6.77)$$

$$x_j^k \geqslant 0 \qquad \text{for} \quad j \in S, \quad k \in K_j \qquad (6.78)$$

This problem is similar to that of Section 2.3. Here, however, we apply a different derivation from that of Section 2.3 using an approach similar to that of Section 2.3, and obtain similar results.

We present only the following theorem, which should prove useful later.

TABLE 6.1

The Tucker Diagram for the Semi-Markovian Decision Process with Discounting

Primal

Variables	$x_1^1 > 0$	$x_1^2 > 0$	$\cdots$	$x_2^1 > 0$	$x_2^2 > 0$	$\cdots$	$x_N^1 > 0$	$x_N^2 > 0$	$\cdots$	Relations	Constants
v_1	$1 - q_{11}^1(\alpha)$	$1 - q_{11}^2(\alpha)$	$\cdots$	$-q_{21}^1(\alpha)$	$-q_{21}^2(\alpha)$	$\cdots$	$-q_{N1}^1(\alpha)$	$-q_{N1}^2(\alpha)$	$\cdots$	$=$	α_1
v_2	$-q_{12}^1(\alpha)$	$-q_{12}^2(\alpha)$	$\cdots$	$1 - q_{22}^1(\alpha)$	$1 - q_{22}^2(\alpha)$	$\cdots$	$-q_{N2}^1(\alpha)$	$-q_{N2}^2(\alpha)$	$\cdots$	$=$	α_2
$\cdots$	$\cdots$	$\cdots$		$\cdots$	$\cdots$		$\cdots$	$\cdots$		$\cdots$	$\cdots$
v_{N-1}	$-q_{1,N-1}^1(\alpha)$	$-q_{1,N-1}^2(\alpha)$	$\cdots$	$-q_{2,N-1}^1(\alpha)$	$-q_{2,N-1}^2(\alpha)$	$\cdots$	$1 - q_{N,N-1}^1(\alpha)$	$-q_{N,N-1}^2(\alpha)$	$\cdots$	$=$	α_{N-1}
v_N	$-q_{1N}^1(\alpha)$	$-q_{1N}^2(\alpha)$	$\cdots$	$-q_{2N}^1(\alpha)$	$-q_{2N}^2(\alpha)$	$\cdots$	$1 - q_{NN}^1(\alpha)$	$1 - q_{NN}^2(\alpha)$	$\cdots$	$=$	α_N
Relations	$\wedge$	$\wedge$		$\wedge$	$\wedge$		$\wedge$	$\wedge$	$\cdots$		
Constants	$\rho_1^1(\alpha)$	$\rho_1^2(\alpha)$		$\rho_2^1(\alpha)$	$\rho_2^2(\alpha)$	$\cdots$	$\rho_N^1(\alpha)$	$\rho_N^2(\alpha)$	$\cdots$		

Dual

THEOREM 6.3.   *For all positive right-hand side $a_j > 0$ (say $a_j = 1/N$), there exists a basic feasible solution with property such that for each $i \in S$ there is only one k such that $x_i^k > 0$ and $x_i^k = 0$ for k otherwise.*

Since the proof is essentially the same as that of Corollary 2.11, we omit it here.

This theorem, which asserts that any basic feasible solution forms a non-randomized stationary strategy, plays an important role in the policy iteration algorithm.

We next consider the dual problem of the primal problem (6.76) through (6.78) (since a discussion similar to that of Section 2.3 implies that the constraints (6.77) have rank N and that the dual problem has meaning). The *dual problem* is:

$$\min \sum_{i \in S} a_i v_i \tag{6.79}$$

subject to

$$v_i \geqslant \rho_i^k(\alpha) + \sum_{j \in S} q_{ij}^k(\alpha) v_j \qquad \text{for} \quad i \in S, \quad k \in K_i \tag{6.80}$$

$$v_i, \qquad \text{unconstrained in sign for} \quad i \in S \tag{6.81}$$

where the dual variable v_i ($i \in S$) is the ith element of $v(\alpha, \pi^*)$, where π^* is an α-optimal strategy.

Table 6.1 shows the Tucker diagram of the primal and dual problems. The relation between policy iteration and linear programming algorithms has been stated in Section 2.4. For this problem we can apply a similar approach. Using the results of Section 2.4 for this problem and Theorem 6.3 of the primal problem, we have immediately the policy iteration algorithm for the semi-Markovian decision process with discounting. We first note that the dual variables in (6.81) are also the simplex multipliers for the primal problem.

The policy iteration algorithm consists of the following two parts:

Value Determination Operation

Take any stationary strategy f^{∞}. Solve

$$v_i = \rho_i^k(\alpha) + \sum_{j \in S} q_{ij}^k(\alpha) v_j$$

for v_i ($i \in S$), where the superscript k corresponds to the chosen strategy f^{∞}.

Policy Improvement Routine

Using the values v_i ($i \in S$), find the element of $G(i, f)$ for each $i \in S$ such that

$$\rho_i^k(\alpha) + \sum_{j \in S} q_{ij}^k(\alpha) v_j > v_i$$

for all $k \in K_i$. If $G(i, f)$ is empty for all $i \in S$, f^{∞} is α-optimal and $v(\alpha, f^{\infty}) = [v_i]$ is the discounted total expected return. If at least $g(i) \in G(i, f)$ for some i, make an improved strategy g^{∞} such that $g(i) \in G(i, f)$ for some i and $g(i) = f(i)$ for $G(i, f)$ empty, and return to the value determination operation.

Here we use the same notation f^{∞} used to denote a stationary strategy in the discrete time case.

As an initial strategy f^{∞}, we may take, for example, $\max_{k \in K_i} \rho_i{}^k(\alpha)$ for each $i \in S$. The reason for this selection can be examined from the linear programming viewpoint.

According to linear programming, the policy improvement routine is an extension of the simplex criterion such that the simplex criteria for each state are applied simultaneously. If $g(i) \in G(i, f)$ for some one i, and $G(i, f)$ is empty for other i, the policy improvement routine is equivalent to the simplex criterion of the primal problem.

We shall show that an improved strategy g^{∞} from the policy improvement routine satisfies $v(\alpha, g^{\infty}) > v(\alpha, f^{\infty})$.

THEOREM 6.4. *Take any f^{∞}. If $g(i) \in G(i, f)$ for some i and otherwise $g(i) = f(i)$ for $G(i, f)$ empty, then $v(\alpha, g^{\infty}) > v(\alpha, f^{\infty})$.*

PROOF. For two strategies f^{∞} and g^{∞}, we have

$$v(\alpha, f^{\infty}) = \rho^f(\alpha) + q^f(\alpha)v(\alpha, f^{\infty}) \qquad (6.82)$$

$$v(\alpha, g^{\infty}) = \rho^g(\alpha) + q^g(\alpha)v(\alpha, g^{\infty}) \qquad (6.83)$$

Subtracting (6. 82) from (6.83), we have

$$v(\alpha, g^{\infty}) - v(\alpha, f^{\infty}) = \rho^g(\alpha) - \rho^f(\alpha) + q^g(\alpha)v(\alpha, g^{\infty}) - q^f(\alpha)v(\alpha, f^{\infty}) \qquad (6.84)$$

We define the $N \times 1$ column vector

$$\gamma = [\gamma_i] = \rho^g(\alpha) + q^g(\alpha)v(\alpha, f^{\infty}) - \rho^f(\alpha) - q^f(\alpha)v(\alpha, f^{\infty}) \qquad (6.85)$$

where $\gamma_i > 0$ if $g(i) \in G(i, f)$ for some i and $\gamma_i = 0$ if $G(i, f)$ is empty from the hypothesis of the theorem. Combining (6.84) and (6.85), and defining $\Delta v = v(\alpha, g^{\infty}) - v(\alpha, f^{\infty})$, we have

$$\Delta v = \gamma + q^g(\alpha)\Delta v \qquad (6.86)$$

Solving for Δv, we obtain

$$\Delta v = [I - q^g(\alpha)]^{-1}\gamma \qquad (6.87)$$

Since (6.85) and $[I - q^g(\alpha)]^{-1}$ are both nonnegative and both never zero simultaneously, we have that at least one element of Δv is positive, that is, $\Delta v > 0$. ∎

We have seen from Theorem 6.4 that the policy iteration algorithm terminates an α-optimal stationary strategy with finite interations, since the number of stationary strategies is finite. Note that if there are two or more strategies satisfying the optimality equation

$$v(\alpha, f^{\infty}) = \max_{f} \left[\rho^f(\alpha) + q^f(\alpha)v(\alpha, f^{\infty}) \right]$$

then these strategies are all α-optimal.

6.5. Semi-Markovian Decision Processes with No Discounting

We shall now develop the semi-Markovian decision processes with no discounting. For the discounted process we have seen that there is an α-optimal stationary strategy. For the nondiscounted process we shall first prove the same result.

As a criterion for the nondiscounted model, we define for any strategy π

$$u(\pi) = \liminf_{t \to \infty} (v(t; \pi)/t) \tag{6.88}$$

where $v(t, \pi)$ is the total expected return ($N \times 1$ column vector) up to time t starting in each state using a strategy π. Then $u(\pi)$ is the long-run average return per unit time, which corresponds to (3.12) in the discrete time case. If we take a stationary strategy, we obtain $v(t; \pi)$ by solving (6.48) for $v_i(t)$. But for any nonstationary strategy, we cannot write it explicitly.

Our problem then is to find a strategy which maximizes (6.88) under all strategies π, that is, an *optimal strategy* π^* such that $u(\pi^*) \geqslant u(\pi)$ for all π.

THEOREM 6.5.   *There is an optimal strategy that is stationary.*

PROOF.   We have seen from Theorem 6.2 that there is an α-optimal stationary strategy. Since the number of stationary strategies is finite, it is possible to choose a sequence $\{\alpha_\nu\}$, $\lim_{\nu \to \infty} \alpha_\nu = 0$, such that $\pi^* = \pi_{\alpha_\nu}$, $\nu = 1, 2, \ldots$, where $\pi_{\alpha\nu}$ is a stationary strategy. We have from (6.47) that

$$v(\alpha, f) = u(f)/\alpha + v(f) + 0(1) \tag{6.89}$$

for any stationary strategy f^{∞}. Thus using Abelian and Tauberian theorems (Widder [119]), we have

$$u(\pi) \leqslant \liminf_{\nu \to \infty} \alpha_\nu v(\alpha_\nu, \pi) \leqslant \liminf_{\nu \to \infty} \alpha_\nu v(\alpha_\nu, \pi_{\alpha_\nu}) = u(\pi^*) \tag{6.90}$$

for any strategy π, which shows that π^* is an optimal stationary strategy. ∎

Applying Theorem 6.5, we may find optimal strategies within the stationary ones under the average criterion. Thus we restrict our attention to stationary strategies and we can then immediately apply the results of Section 6.3.

In the remaining part of this section we shall discuss the linear programming formulation for the completely ergodic process and the terminating process. Further, we can similarly obtain the corresponding policy iteration algorithms from the straightforward results of Chapter 3.

We shall consider the *completely ergodic* case. For this problem the long-run average return per unit time is

$$g = \sum_{j \in S} \pi_j \eta_j r_j / \sum_{j \in S} \pi_j \eta_j \tag{6.91}$$

where π_j is the limiting probability in state j of the imbedded Markov chain. Now we shall consider the decision process. The objective function is

$$\sum_{j \in S} \pi_j(f) \eta_j^k r_j^k / \sum_{j \in S} \pi_j(f) \eta_j^k \tag{6.92}$$

where $\pi_j(f)$ is the corresponding limiting probability of a stationary strategy f^∞, and η_i^k and r_i^k are the corresponding unconditional mean, the return of the strategy f^∞, respectively. The limiting probabilities satisfy

$$\pi_j(f) = \sum_{i \in S} \pi_i(f) p_{ij}(f) \qquad \text{for} \quad j \in S \tag{6.93}$$

$$\sum_{j \in S} \pi_j(f) = 1, \tag{6.94}$$

$$\pi_j(f) > 0 \qquad \text{for} \quad j \in S \tag{6.95}$$

Now let d_j^k ($j \in S$, $k \in K_j$) be the joint probability that the system is in state j and that the decision k is made. It is evident that

$$\sum_{k \in K_j} d_j^k = 1, \qquad 0 \leqslant d_j^k (\leqslant 1) \qquad \text{for} \quad j \in S, \quad k \in K_j \tag{6.96}$$

The objective function of our problem is

$$\sum_{j \in S} \sum_{k \in K_j} \pi_j(f) \eta_j^k r_j^k d_j^k / \sum_{j \in S} \sum_{k \in K_j} \pi_j(f) \eta_j^k d_j^k \tag{6.97}$$

using d_j^k. While (6.93), (6.94), and (6.95) become

$$\pi_j(f) - \sum_{i \in S} \sum_{k \in K_i} \pi_i(f) p_{ij}^k d_i^k = 0 \qquad \text{for} \quad j \in S \tag{6.98}$$

$$\sum_{j \in S} \pi_j(f) = 1 \tag{6.99}$$

$$\pi_j(f) > 0 \qquad \text{for} \quad j \in S \tag{6.100}$$

Setting

$$x_j^k = \pi_j(f) d_j^k \geqslant 0 \qquad \text{for} \quad j \in S, \quad k \in K_j \tag{6.101}$$

and using the fact that $\pi_j(f) = \sum_{k \in K_j} x_j^k$ for $j \in S$, we have from (6.97) through (6.101) the following programming problem:

$$\max \sum_{j \in S} \sum_{k \in K_j} \eta_j^k r_j^k x_j^k / \sum_{j \in S} \sum_{k \in K_j} \eta_j^k x_j^k \tag{6.102}$$

subject to

$$\sum_{k \in K_j} x_j^k - \sum_{i \in S} \sum_{k \in K_i} p_{ij}^k x_i^k = 0 \qquad \text{for} \quad j \in S \tag{6.103}$$

$$\sum_{j \in S} \sum_{k \in K_j} x_j^k = 1 \tag{6.104}$$

$$x_j^k \geqslant 0 \qquad \text{for} \quad j \in S, \quad k \in K_j \tag{6.105}$$

This problem is a so-called linear fractional programming problem (see, for example, Charnes and Cooper [21]) since the objective function (6.102) is linear fractional and the constraints are linear. We can reduce it to a linear programming problem by using a transformation of variables. Noting that $\sum_{j \in S} \sum_{k \in K_j} \eta_j^k x_j^k > 0$, we transform the variables x_j^k to y_j^k and y as follows

$$y_j^k = x_j^k / \sum_{j \in S} \sum_{k \in K_j} \eta_j^k x_j^k \tag{6.106}$$

and

$$y = 1 / \sum_{j \in S} \sum_{k \in K_j} \eta_j^k x_j^k \tag{6.107}$$

Then we have the following linear programming problem:

$$\max \sum_{j \in S} \sum_{k \in K_j} \eta_j^k r_j^k y_j^k \tag{6.108}$$

subject to

$$\sum_{k \in K_j} y_j^k - \sum_{i \in S} \sum_{k \in K_i} p_{ij}^k y_i^k = 0 \qquad \text{for} \quad j \in S \tag{6.109}$$

$$\sum_{j \in S} \sum_{k \in K_j} y_j^k = y \tag{6.110}$$

$$\sum_{j \in S} \sum_{k \in K_j} \eta_j^k y_j^k = 1 \tag{6.111}$$

$$y_j^k \geqslant 0 \qquad \text{for} \quad j \in S, \quad k \in K_j \tag{6.112}$$

where the additional constraint (6.111) is derived from the transformations of (6.106) and (6.107).

We shall next consider some properties of the above linear programming problem.

THEOREM 6.6. *For the linear programming problem (6.108) through (6.112), there is a basic feasible solution with the property that, for each $i \in S$, there is only one k such that $y_i^k > 0$ and $y_i^k = 0$ for k otherwise.*

The proof of this theorem is similar to that of Theorem 3.8, so we omit it here. Theorem 6.6 implies that any basic feasible solution always has $y > 0$.

THEOREM 6.7. *The semi-Markovian decision process under consideration is equivalent to the linear programming problem:*

$$\max \sum_{j \in S} \sum_{k \in K_j} \eta_j{}^k r_j{}^k y_j{}^k \tag{6.113}$$

subject to

$$\sum_{k \in K_j} y_j{}^k - \sum_{i \in S} \sum_{k \in K_i} p_{ij}^k y_i{}^k = 0 \quad \text{for} \quad j = 1, \ldots, N-1 \tag{6.114}$$

$$\sum_{j \in S} \sum_{k \in K_j} \eta_j{}^k y_j{}^k = 1 \tag{6.115}$$

$$y_j{}^k \geqslant 0 \quad \text{for} \quad j \in S, \quad k \in K_j \tag{6.116}$$

That is, we need not to take account of the constraint (6.110) *and of the variable y. Here we omit the redundant constraint for* $j = N$ *in* (6.114).

PROOF.   As we have noted above, any basic feasible solution always has $y > 0$. We require however the optimal stationary strategy and its associated maximum average return per unit time g. It is clear that the optimal value of (6.113) has no relation to y. From the transformation of variables in (6.106) and (6.107), we have $x_j{}^k = y y_j{}^k$ for $j \in S$, $k \in K_j$. Thus the optimal stationary strategy is determined by

$$d_j{}^k = x_j{}^k / \sum_{k \in K_j} x_j{}^k = y y_j{}^k / \sum_{k \in K_j} y y_j{}^k = y_j{}^k / \sum_{k \in K_j} y_j{}^k \tag{6.117}$$

which is independent of y. ∎

Theorem 6.7 gives the linear programming problem (6.113) through (6.116) which has $\sum_{i \in S} K_i$ variables and N constraints, and where the redundant constraint for $j = N$ in (6.114) is omitted.

Now we shall consider the dual problem, where we consider the problem of (6.113) through (6.116) as a primal problem. Defining dual variables $(v_1, v_2, \ldots, v_{N-1}, g)$, we have the dual problem:

$$\min g \tag{6.118}$$

subject to

$$\eta_i{}^k g + v_i \geqslant \eta_i{}^k r_i{}^k + \sum_{j=1}^{N-1} p_{ij}^k v_j \quad \text{for} \quad j \in S, \quad k \in K_j \tag{6.119}$$

$$g, v_i, \quad \text{unconstrained in sign for} \quad i = 1, 2, \ldots, N-1 \tag{6.120}$$

where we suppose $v_N = 0$ in (6.119). Here we have used the known fact that the constraints (6.114) and (6.115) have rank N and the dual problem has meaning. Further we have also used the fact that the objective function of the dual problem is g, the average return per unit time, from the *duality theorem* (see, for example, Dantzig [25]). Table 6.2 shows the Tucker diagram of the primal and dual problems for the completely ergodic semi-Markovian decision process.

We see from a discussion similar to that of Section 3.4 that the dual variables v_i $(i = 1, 2, \ldots, N - 1)$ are relative bias terms, which correspond to (6.46). For this problem we can obtain the policy iteration algorithm from the similar discussion of Section 3.5. We note that the *dual variables* $(v_1, v_2, \ldots, v_{N-1}, g)$ are also the *simplex multipliers*. From the linear programming viewpoint we have for any basic variables

$$\Delta_i^* = -\eta_i^* r_i^* - \sum_{j=1}^{N-1} p_{ij}^* v_j + \eta_i^* g + v_i = 0 \qquad \text{for} \quad i \in S \qquad (6.121)$$

where the asterisk * denotes the data of any basic feasible solution (and also any stationary strategy). If

$$\Delta_i^k = -\eta_i^k r_i^k - \sum_{j=1}^{N-1} p_{ij}^k v_j + \eta_i^k g + v_i < 0 \qquad (6.122)$$

for some $i \in S$ and $k \in K_i$, we can obtain an improved strategy. Further, if $\Delta_i^k \geqslant 0$ for all $i \in S$ and $k \in K_i$, we have an optimal strategy. Solving for g in (6.122), we have

$$g < r_i^k + \frac{1}{\eta_i^k} \left[\sum_{j=1}^{N-1} p_{ij}^k v_j - v_i \right] \qquad (6.123)$$

which implies the policy improvement routine in the policy iteration algorithm. Thus we have the following policy iteration algorithm.

Value Determination Operation

Take any stationary f^∞. Solve

$$\eta_i^k g + v_i = \eta_i^k r_i^k + \sum_{j=1}^{N-1} p_{ij}^k v_j$$

for $g, v_1, v_2, \ldots, v_{N-1}$ (setting $v_N = 0$), where the superscript k is determined by the chosen strategy f^∞.

Policy Improvement Routine

Using the values v_i, find the element of $G(i, f)$ for each $i \in S$ such that

$$r_i^k + (\eta_i^k)^{-1} \left[\sum_{j=1}^{N-1} p_{ij}^k v_j - v_i \right] > g$$

for all $k \in K_i$. If $G(i, f)$ is empty for all $i \in S, f^\infty$ is optimal, and g is the average return per unit time; then $v_1, \ldots, v_{N-1}$ are the relative bias terms. If $g(i) \in G(i, f)$ for some i, use an improved strategy such that $g(i) \in G(i, f)$, $g(i) = f(i)$ for $G(i, f)$ empty, and return to the value determination operation.

TABLE 6.2

The Tucker Diagram for the Completely Ergodic Semi-Markovian Decision Process

		Primal										
Variables	$y_1^1 \geq 0$	$y_1^2 \geq 0$	⋯	$y_2^1 \geq 0$	$y_2^2 \geq 0$	⋯	$y_N^1 \geq 0$	$y_N^2 \geq 0$	⋯	Relations	Constants
v_1	$1-p_{11}^1$	$1-p_{11}^2$	⋯	$-p_{21}^1$	$-p_{21}^2$	⋯	$-p_{N1}^1$	$-p_{N1}^2$	⋯	$=$	0
v_2	$-p_{12}^1$	$-p_{12}^2$	⋯	$1-p_{22}^1$	$1-p_{22}^2$	⋯	$-p_{N2}^1$	$-p_{N2}^2$	⋯	$=$	0
⋮	⋮	⋮		⋮	⋮		⋮	⋮		⋮	⋮
v_{N-1}	$-p_{1,N-1}^1$	$-p_{1,N-1}^2$	⋯	$-p_{2,N-1}^1$	$-p_{2,N-1}^2$	⋯	$-p_{N,N-1}^1$	$-p_{N,N-1}^2$	⋯	$=$	0
v_N	η_1^1	η_1^2	⋯	η_2^1	η_2^2	⋯	η_N^1	η_N^2	⋯	$=$	1
Relations	$\wedge$	$\wedge$	⋯	$\wedge$	$\wedge$	⋯	$\wedge$	$\wedge$	⋯		
Constants	$\eta_1^1 r_1^1$	$\eta_1^2 r_1^2$	⋯	$\eta_2^1 r_2^1$	$\eta_2^2 r_2^2$	⋯	$\eta_N^1 r_N^1$	$\eta_N^2 r_N^2$	⋯		

Dual

We shall show only that the policy improvement routine yields the improved strategy whose average return is greater than the previous one.

THEOREM 6.8. *Take any f^∞. If $g(i) \in G(i, f)$ for some i and otherwise $g(i) = f(i)$ for $G(i, f)$ empty, then $g(g) > g(f)$, where $g(f)$ denotes the average return per unit time using a strategy f^∞.*

PROOF.   For any two strategies f^∞ and g^∞, we have

$$\eta_i^f g(f) + v_i(f) = \eta_i^f r_i^f + \sum_{j=1}^{N-1} p_{ij}^f v_j(f) \tag{6.124}$$

$$\eta_i^g g(g) + v_i(g) = \eta_i^g r_i^g + \sum_{j=1}^{N-1} p_{ij}^g v_j(g) \tag{6.125}$$

where η_i^f and p_{ij}^f denote η_i^k and p_{ij}^k, respectively, where $k = f(i)$. Dividing (6.124), (6.125) by η_i^f and η_i^g, respectively, and subtracting the former from the latter, we obtain

$$v_i(g)/\eta_i^g - v_i(f)/\eta_i^f + g(g) - g(f) = r_i^g - r_i^f + (\eta_i^g)^{-1} \sum_{j=1}^{N-1} p_{ij}^g v_j(g)$$

$$- (\eta_i^f)^{-1} \sum_{j=1}^{N-1} p_{ij}^f v_j(f) \tag{6.126}$$

We define the $N \times 1$ column vector

$$\gamma = [\gamma_i] = \left[r_i^g - r_i^f + (\eta_i^g)^{-1} \left\{ \sum_{j=1}^{N-1} p_{ij}^g v_i(g) - v_i(g) \right\} \right.$$

$$\left. + (\eta_i^f)^{-1} \left\{ \sum_{j=1}^{N-1} p_{ij}^f v_j(f) - v_i(f) \right\} \right] \tag{6.127}$$

where $\gamma_i > 0$ if $g(i) \in G(i, f)$ for some i and $\gamma_i = 0$ if $G(i, f)$ is empty from the hypothesis of the theorem. Note that at least one $\gamma_i > 0$ for some i. Combining (6.126) and (6.127) and defining

$$\Delta v_i = v_i(g) - v_i(f) \qquad (i = 1, \ldots, N - 1)$$
$$\Delta g = g(g) - g(f)$$

we have

$$\Delta v_i + \eta_i^g \Delta g = \eta_i^g \gamma_i + \sum_{j=1}^{N-1} p_{ij}^g \Delta v_j \tag{6.128}$$

Let $\pi(g) = (\pi_1(g), \ldots, \pi_N(g))$ be the limiting probability vector of the imbedded Markov chain $P(g) = [p_{ij}^g]$. It is evident that $\pi(g) = \pi(g)P(g)$. Premultiplying (6.128) by $\pi_i(g)$ and summing on all $i \in S$, we have

$$\Delta g = \sum_{i \in S} \pi_i(g)\eta_i^g \gamma_i / \sum_{i \in S} \pi_i(g)\eta_i^g \tag{6.129}$$

which is positive since the limiting probability in state i is

$$\pi_i(g)\eta_i{}^g / \sum_{i \in S} \pi_i(g)\eta_i{}^g > 0$$

for all $i \in S$, and at least one $\gamma_i > 0$. Thus we have

$$\Delta g = g(g) - g(f) > 0. \quad \blacksquare \qquad (6.130)$$

Theorem 6.8 implies that the policy iteration algorithm terminates an optimal stationary strategy with finite iterations. Note that if there are two or more strategies satisfying the *optimality equation*

$$\eta_i{}^k g + v_i = \max_k \left[\eta_i{}^k r_i{}^k + \sum_{j=1}^{N-1} p_{ij}^k v_j \right] \qquad (i \in S)$$

then these strategies are all optimal and have the same average return g.

In this case we encounter the problem of finding a strategy having the maximal bias term among the optimal strategies. That is, we should require a 0-*optimal* strategy which refers to a 1-optimal strategy in the discrete time model. We omit the problem for finding 0-optimal strategies here.

Finally we shall consider the *terminating process*. For the terminating process, the total expected return before absorption starting in an initial distribution a is

$$a'v' = a'[I - P_1]^{-1} r' \qquad (6.131)$$

where we define that state 1 is absorbing. For the semi-Markovian decision process with which we are concerned, we shall first provide the following assumption (cf. Section 3.7):

Terminating assumption. The common absorbing state can be reached with probability 1 from any transient state in a finite time whatever decisions are made.

Under the above assumption our problem is to find a strategy having the maximal total expected return before absorption among all strategies. Thus we need not to consider the decision in state 1 since our concern is the behavior before absorption. This problem is very similar to that of the discounted model. Hence, we shall only give the following linear programming problem:

$$\max \sum_{j=2}^{N} \sum_{k \in K_j} r_j{}^k \eta_j{}^k x_j{}^k \qquad (6.132)$$

subject to

$$\sum_{k \in K_j} x_j{}^k - \sum_{i \in S} \sum_{k \in K_i} p_{ij}^k x_i{}^k = a_j \qquad \text{for} \quad j = 2, \ldots, N \qquad (6.133)$$

$$x_j{}^k \geqslant 0 \qquad \text{for} \quad j = 2, \ldots, N, \quad k \in K_j \qquad (6.134)$$

The next theorem is obvious.

THEOREM 6.9.   *For all positive right-hand side $a_j > 0$ [say, $a_j = 1/(N - 1)$], there exists any basic feasible solution with property such that for each $i = 2, \ldots, N$, there is only one k such that $x_i^k > 0$ and $x_i^k = 0$ for k otherwise.*

The above linear programming problem and Theorem 6.9 immediately imply the corresponding policy iteration algorithm.

References and Comments

For the theory of semi-Markov process or Markov renewal processes, see Smith [107], Pyke [97, 98], Barlow and Proschan [5], and others. For renewal theory, see Cox [24] and Smith [108].

Semi-Markovian decision processes, or Markov renewal programs were first presented in 1963 by Jewell [65, 66]. Howard [64] and De Cani [26] have also studied policy iteration algorithms for the processes. Optimality of stationary strategies has been given by Denardo [31] † for the discounted model, and by Fox [59] ‡ for the nondiscounted model. Further, Fox [122] has shown the optimality of stationary strategies for the nondiscounted model under the mild condition. Linear programming formulations of the processes have been given by Howard [64], Fox [59], De Ghellinck and Eppen [28, appendix], and Osaki and Mine [94]. Further Denardo and Fox [33] have studied the general multichain semi-Markovian decision processes.

The decision process on a continuous time Markov process has first given by Howard [63], and Rykov [103] has shown the optimality of stationary strategies for the discounted model. This is not discussed in this chapter. Miller [89, 90] has given the strengthened results for the processes over a finite or an infinite time horizon. Martin–Löf [88] has shown that there exists a periodic optimal strategy for a continuous-time Markov process with periodic transition probability under the total expected return for the discounted model.

Chapter 7

Generalized Markovian Decision Processes

7.1. Introduction

In the preceding chapter we discussed semi-Markovian decision processes in which transition time from state i to state j is a random variable. In this chapter we return to the simple models in which the decisions are made periodically and there is a discount factor β ($0 \leqslant \beta < 1$). Here, however, we generalize state and action spaces to any nonempty Borel sets. The model discussed in this chapter is a generalization of that of Chapter 2. Thus special cases—for example, the models whose state and action spaces are both finite sets—reduce to those of the preceding discussion.

In Section 7.2, we introduce the probabilistic definitions and notation to prepare for later discussion, and, in Section 7.3, we state the existence theorems for the generalized Markovian decision processes with discounting. In the final section of this chapter we discuss special cases.

7.2. Definitions and Notation

Before introducing generalized Markovian decision processes, we give some probabilistic definitions and notation.

By a Borel set we mean a Borel subset of some complete separable metric space. A probability on a nonempty Borel set X is a probability measure defined over the Borel subsets of X; the set of all probabilities on X is denoted by $P(X)$. For any nonempty Borel sets X, Y, a conditional probability on Y given X is a function $q(\mid)$ such that for each $x \in X$, $q(\mid x)$ is a probability on Y and for each Borel set $B \subset Y$, $q(B \mid)$ is a Baire function on X. The set of all conditional probabilities on Y given X is denoted by $Q(Y|X)$. The product space of X and Y will be denoted by XY. The set of bounded Baire functions on X is denoted by $M(X)$.

For any $u \in M(XY)$ and any $q \in Q(Y|X)$, qu denotes the element of $M(X)$ whose value at $x_0 \in X$ is

$$qu(x_0) = \int_Y u(x_0, y)\, dq(y|x_0) \tag{7.1}$$

For any $p \in P(X)$ and any $u \in M(X)$, pu is the integral of u with respect to p, that is,

$$pu = \int_X u(x)\, dp(x) \tag{7.2}$$

For any $p \in P(X)$, $q \in Q(Y|X)$, pq is the probability on XY such that, for every $u \in M(XY)$, $pq(u) = p(qu)$. Conversely, we consider every probability m on XY has a factorization $m = pq$, where p is unique and is the marginal distribution of the first coordinate variable with respect to m; q is not unique. It is a version of the conditional distribution of the second coordinate variable given the first. These facts will be found in the texts on probability theory.

We now extend the above notation in an obvious way to a finite or countable sequence of nonempty Borel sets $X_1, X_2, \ldots$. If $q_n \in Q(X_{n+1}|X_1 \cdots X_n)$ for $n \geqslant 1$ and $p \in P(X_1)$, $pq_1 \cdots q_n$ is a probability on $X_1 X_2 \cdots X_{n+1}$. Further, $pq_1 q_2 \cdots$ is a probability on the infinite product space $X_1 X_2 \cdots$, $q_2 q_3 \in Q(X_3 X_4 | X_1 X_2)$, for any $u \in M(X_1 X_2 \cdots X_{n+1})$, $n \geqslant 1$ and any m, $1 \leqslant m \leqslant n$, $q_m \cdots q_n u \in M(X_1 X_2 \cdots X_m)$, etc.

To avoid further complicating the notation which was already involved, we introduce an ambiguity as follows; for any function u on Y, we shall use the same symbol u to denote the function v on XY such that $v(x, y) = u(y)$ for all y. Thus, for example, for any $q \in Q(Y|X)$, $u \in M(Y)$, $qu \in M(X)$; any $q \in Q(Y|X)$ will also denote the element q' of $Q(Y|ZX)$ defined by $q'(\ |z,) = q(\ | \ )$, etc.

A $p \in P(X)$ is *degenerate* if it is concentrated at some one point $x \in X$; a $q \in Q(Y|X)$ is degenerate if each $q(\ |x)$ is degenerate. The degenerate q are exactly those for which there is a Baire function f mapping X into Y for which $q(\{f(x)\}|x) = 1$ for all $x \in X$. Any such f will also denote its associated degenerate q, so that, for any $u \in M(XY)$, $fu(x) = u(x, f(x))$ for all $x \in X$.

A generalized Markovian decision process with discounting is defined by the quintuplet (S, A, q, r, β), where state space S and action space A are any nonempty Borel sets of some complete separable metric space, $q \in Q(S|SA)$, $r \in M(SAS)$ and β ($0 \leqslant \beta < 1$). In our problem we make the decisions at the discrete time $n = 1, 2, \ldots$, that is, when the system is in $s \in S$ and we make the decision $a \in A$, the system obeys the probability law $q(\ |s, a)$ on S and we receive the return $r(s, a,)$. The quintuplet (S, A, q, r, β) corresponds to that of Chapter 2, where $r \in M(SAS)$ refers to r_{ij}^k in Section 2.5. Since we have already considered strategies, let us now define *plans* for the model considered.

DEFINITION 7.1. *A plan* π *is a sequence* $(\pi_1, \pi_2, \ldots)$, *where* $\pi_n \in Q(A|H_n)$ *and* $H_n = SA \cdots S$ ($2n - 1$ *factors*) *is the set of possible histories of the system when the decision at time n is made.*

A plan π depends on the history of all states and decisions.

DEFINITION 7.2. *A plan π is* randomized Markov *if $\pi_n \in Q(A|S)$ for each $n = 1, 2, \ldots$.*

DEFINITION 7.3. *A plan π is* nonrandomized Markov *if the plan π is Markov and each π_n is degenerate, that is, $\pi = (f_1, f_2, \ldots)$, where each f_n is a Baire function from S into A.*

A strategy we have discussed in the preceding chapters (except for Chapter 6) corresponds to a nonrandomized Markov plan and a randomized strategy corresponds to a randomized Markov plan.

DEFINITION 7.4. *A plan π is* randomized stationary *if the plan π is Markov and $\pi_n = \pi_1$ for all $n \geqslant 2$.*

DEFINITION 7.5. *A plan π is* nonrandomized stationary *if the plan π is stationary and each π_n is degenerate, that is, $\pi = (f, f, \ldots)$, where f is a Baire function from S into A. The nonrandomized stationary plan defined by f is denoted by f^∞.*

We shall give the following lemma which is useful in the later discussion.

LEMMA 7.6. *For any $q \in Q(Y|X)$, $u \in M(XY)$, $\varepsilon > 0$, there is a degenerate $f \in Q(Y|X)$ such that*

$$fu \geqslant qu \qquad \text{for all} \quad x \in X \tag{7.3}$$

and

$$q(\{y: u(x_0, y) \geqslant (x_0, f(x_0)) + \varepsilon\}|x_0) = 0 \tag{7.4}$$

for all $x_0 \in X$.

We will omit the proof here, but it may be found in Blackwell [15].

The lemma states that, in the situation where we observe $x \in X$, then choose $y \in Y$, receiving the return $u(x, y)$, any randomized plan can be replaced by a nonrandomized plan f such that (7.3) the expected return for each x, is at least as large as it was before and (7.4) with probability 1, for each x, the actual return under q does not exceed the actual return under f by as much as ε.

We now find the discounted total expected return for the process. Any plan π, together with the law of motion q of the system, defines for each initial state s a conditional distribution on the set $\Omega = ASAS \cdots$ of futures of the system, that is, it defines an element of $Q(\Omega|S)$, namely

$$e_\pi = \pi_1 q_1 \pi_2 q_2 \cdots$$

where

$$q_n = q(\ |s_n, a_n) \in Q(S|SA) \subset Q(S|H_nA) \tag{7.5}$$

8

for $n = 1, 2, \ldots$. Let $V(\pi)(s_1)$ be the discounted total expected return starting in state s_1 and using a plan π. Denote the coordinate function $S\Omega$ by $s_1, a_1, s_2, a_2, \ldots$, so that the return at time n, as a function of the history of the system, is $r(s_n, a_n, s_{n+1})$, and the discounted total return is

$$u = \sum_{n=1}^{\infty} \beta^{n-1} r(s_n, a_n, s_{n-1})$$

Then

$$V(\pi)(s_1) = e_\pi u$$

$$= \int_{AS} r(s_1, a_1, s_2)\, d\pi_1(a_1|s_1)\, dq(s_2|s_1, a_1)$$

$$+ \beta \int_{ASAS} r(s_2, a_2, s_3)\, d\pi_1(a_1|s_1)\, dq(s_2|s_1, a_1)\, d\pi_2(a_2|s_1, a_1, s_2) + \cdots$$

$$= \sum_{n=1}^{\infty} \beta^{n-1} \pi_1 q \cdots \pi_n q r \tag{7.6}$$

where $\pi_n \in Q(A|H_n)$, $q = q(s_{n+1}|s_n, a_n) \in Q(S|SA) \subset Q(S|H_nA)$,

$$\pi_1 q \cdots \pi_n q \in Q(H_{n+1}|S)$$

and

$$r = r(s_n, a_n, s_{n+1}) \in M(SAS) \subset M(H_{n+1})$$

Note that we use the ambiguous notation in (7.6).

For the generalized Markovian decision process, we shall show the existence theorems of optimal plans that maximize (7.6) over all π. Here we define optimal plans as follows.

DEFINITION 7.7.   *For any $p \in P(S)$, and any $\varepsilon > 0$, π^* is (p, ε)-optimal if* $p\{s|V(\pi)(s) > V(\pi^*)(s) + \varepsilon\} = 0$ *for every π.*

DEFINITION 7.8.   *For any $\varepsilon > 0$, π^* is ε-optimal if it is (p, ε)-optimal for every p, or, equivalently, if $V(\pi)(s) \leqslant V(\pi^*)(s) + \varepsilon$ for all π, s.*

DEFINITION 7.9.   *For any $p \in P(S)$, π^* is p-optimal if (p, ε)-optimal for every $\varepsilon > 0$, or, equivalently, $p\{s|V(\pi)(s) > V(\pi^*)(s)\} = 0$ for all π.*

DEFINITION 7.10. π^* *is optimal if $V(\pi)(s) \leqslant V(\pi^*)(s)$ for all π, s.*

The definition of optimal plans will coincide with that of β-optimal strategies in Definition 2.1. In the next section we give existence theorems of (p, ε)-optimal, p-optimal, ε-optimal, and optimal plans of this nature.

7.3. Existence Theorems

We shall demonstrate the existence of (p, ε)-optimal π:

THEOREM 7.11.   *For any $p \in P(S)$ and $\varepsilon > 0$ there is a (p, ε)-optimal plan.*

PROOF.   We associate with each π the number $pV(\pi) = \int_S V(\pi)(s)dp(s)$, the expected return from π when the initial state has distribution p, denote by v the upper bound over all π of the numbers $pV(\pi)$, choose a sequence $\pi^{(1)}, \pi^{(2)}, \ldots$ with $pV(\pi^{(n)}) \to v$, and set $u = \sup V(\pi^{(n)})$.

Let S_n consist of all s for which n is the smallest k with $V(\pi^{(k)})(s) \geqslant u - \varepsilon$, and let π^* be the plan which uses $\pi^{(n)}$ for all initial states $s \in S_n$, that is,

$$\pi_m^*(\ |s_1, a_1, \ldots, s_m) = \pi_m^{(n)}(\ |s_1, a_1, \ldots, s_m) \qquad \text{for} \quad s_1 \in S_n \qquad (7.7)$$

Then $V(\pi^*)(s) = V(\pi^{(n)})(s)$ on $s \in S_n$, and $V(\pi^*)(s) \geqslant u - \varepsilon$ everywhere. We show that, for every π, $p\{V(\pi) \leqslant u\} = 1$, which will show that π^* is (p, ε)-optimal. Take any π and any $\gamma > 0$. The construction above, applied to the sequence $\pi, \pi^{(1)}, \pi^{(2)}, \ldots$, yields a π^{**} with $V(\pi^{**}) \geqslant \max(u, V(\pi)) - \gamma$ everywhere. But $pV(\pi^{**}) \leqslant v \leqslant pu$, while $pV(\pi^{**}) \geqslant p \max(u, V(\pi)(s)) - \gamma$, so that $p \max(u, V(\pi)(s)) \leqslant pu + \gamma$. Since γ is any positive number,

$$p \max(u, V(\pi)) \leqslant pu, \quad \text{and} \quad p\{s | V(\pi)(s) \leqslant u\} = 1$$

Thus

$$p\{s | V(\pi)(s) \leqslant V(\pi^*)(s) + \varepsilon\} = 1. \ \blacksquare$$

We know the existence of (p, ε)-optimal π, but we present examples in which there are no p-optimal plans or no ε-optimal plans.

EXAMPLE 1.   (No p-optimal plans.) S has a single element, say 0, and A has countably many elements, say $1, 2, 3, \ldots$. We take

$$r(0, a, 0) = (a - 1)/a.$$

There is no π with $V(\pi)(0) = 1/(1 - \beta)$, but $\sup_\pi V(\pi)(0) = 1/(1 - \beta)$.

EXAMPLE 2.   (No ε-optimal plans.) We take $S = A = $ unit interval $[0, 1]$. The state of the system remains fixed: $(s, a) \to s$, and the return r is 1 or 0 depending on whether (s, a) is in a given Borel subset $B \subset SA$ or not. For any $\pi = (\pi_1, \pi_2, \ldots)$, $\{s: \pi_1 q r > 0\}$ is a Borel subset of the projection D of B on S. For B chosen so that D is not a Borel set there is an $s_0 \in D$ for which $\pi_1 r = 0$, so that $V(\pi^*)(s_0) \leqslant \beta + \beta^2 + \cdots = \beta/(1 - \beta)$. For $(s_0, a_0) \in B$ we choose $f^* \in F$ such that $f^*(s_0) = a_0$ and make $\pi^* = f^{*\infty}$. Then $V(\pi^*)(s_0) = 1 + \beta + \beta^2 + \cdots = 1/(1 - \beta)$. For any ε $(0 < \varepsilon < 1)$, $V(\pi)(s_0) + \varepsilon < 1/(1 - \beta) = V(\pi^*)(s_0)$, which shows that π is not ε-optimal for any $\varepsilon < 1$.

THEOREM 7.12. *For any $p \in P(S)$, $\varepsilon > 0$, and π, there is a randomized Markov π^* that dominates π, that is, $p\{s \mid V(\pi^*)(s) \geqslant V(\pi)(s) - \varepsilon\} = 1$.*

PROOF. We may suppose that π is already Markov from some point on, say for $n > N$, since any two strategies π, π' that agree for the first n times have

$$\|V(\pi)(s) - V(\pi)(s)\| \leqslant \beta^n \|r\|/(1 - \beta)$$

where, for any $u \in M(S)$, $\|u\| = \sup |u(s)|$. We now show that if

$$\pi = (\pi_1, \ldots, \pi_N, f_{N+1}, \ldots)$$

is Markov for $n > N$, then for any $\gamma > 0$ there is an f_N mapping S into A with $p\{V(\pi')(s) \geqslant V(\pi)(s) - \gamma\} = 1$, where $\pi' = (\pi_1, \ldots, \pi_{N-1}, f_N, f_{N+1}, \ldots)$. Using this fact N times, with $\gamma = \varepsilon/N$, we will produce a Markov π^* with (p, ε)-dominates π.

To find f_N, we write

$$V(\pi)(s) = \pi_1 q \cdots \pi_{N-1} q(u + \beta^{N-1} \pi_N q v)$$

where $u(s_1, a_1, \ldots, s_N) = \sum_{k=1}^{N-1} \beta^{k-1} r(s_k, a_k, s_{k+1})$ and

$$v(s_n, a_n, s_{n+1}) = r(s_N, a_N, s_{N+1}) + (\sum_{k=1}^{\infty} \beta^k f_{N+1} q \cdots f_{N+k} q r)(s_N, a_N, s_{N+1})$$

It suffices to find f_N for which

$$p\{s_1 \mid \pi_1 q \cdots \pi_{N-1} q f_N w(s_1) \geqslant \pi_1 q \cdots \pi_{N-1} q \pi_N w(s_1) - \gamma\} = 1 \quad (7.8)$$

where $w(s_N, a_N) = \beta^{N-1} q v \in M(SA)$.

Consider the probability $m = p\pi_1 q \cdots \pi_N$ on $SA \cdots SA$ ($2N$ factors), and denote the coordinate variables by $s_1, a_1, \ldots, s_N, a_N$. For any f_N, $x = \pi_1 q \cdots \pi_{N-1} q f_N w(s_1)$ is a version of $E(w(s_N, a_N) \mid s_0)$. If we choose f_N so that $w(s_N, f(a_N)) \geqslant w(s_N, a_N) - \gamma$ with probability 1, we shall have $x \geqslant y - \gamma$ with probability 1, which is equivalent to (7.8). That such an f_N exists follows at once from Lemma 7.6 with $X = S$, $Y = A$, q a version of the conditional distribution of a_N given s_N, $u = w$, and $\varepsilon = \gamma$. ∎

COROLLARY 7.13. *For any $p \in P(S)$, $\varepsilon > 0$, there is a (p, ε)-optimal nonrandomized Markov π^*.*

PROOF. From Theorem 7.11 there is a $(p, \varepsilon/2)$-optimal π, and from Theorem 7.12 there is a nonrandomized Markov π^* with $(p, \varepsilon/2)$-dominates π. This π^* is (p, ε)-optimal. ∎

Corollary 7.13 states that there is a (p, ε)-optimal nonrandomized Markov π^*. Thus we can focus on nonrandomized Markov plans, and now we introduce the operators associated with nonrandomized Markov plans. (In the sequel we abbreviate and say *Markov plans* instead of nonrandomized Markov plans.)

Associated with each Baire function f mapping S into A is a corresponding operator $L(f)$, mapping $M(S)$ into $M(S)$, defined as follows. For $u \in M(S)$, we define

$$L(f)u = fq(r + \beta u) \tag{7.9}$$

where the u on the right-hand side, considered as a function on SAS, depends on only the last coordinate.

For any $u(s)$, $v(s) \in M(S)$, we may write $u \geq v$ if $u(s) \geq v(s)$ for all $s \in S$. We state the following theorem, which is straightforward. The proof is omitted.

THEOREM 7.14. (a) $L(f)$ *is monotone, that is, $u \leq v$ implies $L(f)u \leq L(f)v$.* (b) *For any constant c, $L(f)(u + c) = L(f)u + \beta c$.* (c) *For any Markov $\pi = (f_1, f_2, \ldots)$, $L(f)v(\pi) = V(f, \pi)$, where (f, π) denotes the Markov plan $(f, f_1, f_2, \ldots)$.*

DEFINITION 7.15. *For any Markov $\pi = (f_1, f_2, \ldots)$ f mapping S into A is π-generated if there is a partition of S into Borel sets $S_1, S_2, \ldots$, such that $f = f_n$ on S, where sets $S_1, S_2, \ldots$, are called a* partition *of S if $S_i \cap S_j$ is empty for $i \neq j$ and $\cup_n S_n = S$.*

DEFINITION 7.16. *A Markov $\pi' = (g_1, g_2, \ldots)$ is π-generated if each g_n is π-generated.*

For any Markov plan, we define the operator U mapping $M(S)$ into $M(S)$ as follows:

$$Uu = \sup_n L(f_n)u \tag{7.10}$$

We shall demonstrate the following theorem for the operator U.

THEOREM 7.17. (a) U *is monotone.* (b) *For any constant c,*
$$U(u + c) = Uu + \beta c$$
(c) *For any $L(f)$ associated with a π-generated f, $L(f)u \leq Uu$.* (d) *For any $u \in M(S)$ and $\varepsilon > 0$, there is a π-generated f whose associated $L(f)$ satisfies $L(f)u \geq Uu - \varepsilon$.*

PROOF. (a) and (c) are straightforward. To prove (b), we have
$$L(f_n)(u + c) = L(f_n)u + \beta c \leq Uu + \beta c \tag{7.11}$$

so that $U(u + c) \leqslant Uu + \beta c$. This inequality, with u replaced by $u + c$, and c by $-c$, yields $Uu \leqslant U(u + c) - \beta c$, establishing (b). To prove (d), let S_n consist of all s for which $L(f_m)u < Uu - \varepsilon$ for $m < n$, $L(f_n)u \geqslant Uu - \varepsilon$, and set $f = f_n$ for $s \in S_n$. Then, for any v, $L(f)v = L(f_n)v$ on S_n, since $S_1, S_2, \ldots,$ are a partition of S. In particular, $L(f)u = L(f_n)u \geqslant Uu - \varepsilon$ on S_n, so $L(f)u \geqslant Uu - \varepsilon$ everywhere. ∎

To justify our informal interpretation of U, note that, for any Markov $\pi' = (g_1, g_2, \ldots)$, the total return from π' with termination at time $n + 1$ with final payment u is

$$V_n(\pi', u)(s) = L(g_1)L(g_2) \cdots L(g_n)u(s) \tag{7.12}$$

If π' is π-generated, $L(g_i)v(s) \leqslant Uv(s)$ for all i, so that $V_n(\pi', u)(s) \leqslant U^n u$. To find π' with $V_n(\pi', u)(s) \geqslant U^n u - \varepsilon$, choose any positive numbers ε_i and choose g_i π-generated so that

$$L(g_i)U^{n-i}u(s) \geqslant UU^{n-i}u(s) - \varepsilon_i = U^{n-i+1}u(s) - \varepsilon_i \tag{7.13}$$

By induction downward on i, starting at $i = n$, we obtain

$$L(g_i) \cdots L(g_n)u(s) \geqslant U^{n-i+1}u(s) - d_i \tag{7.14}$$

where $d_i = \varepsilon_i + \beta\varepsilon_{i+1} + \cdots + \beta^{n-i}\varepsilon_n$. For $i = 1$ we obtain

$$V_n(\pi', u)(s) \geqslant U^n u(s) - d_1,$$

and the ε_i can be chosen so that $d_1 \leqslant \varepsilon$.

THEOREM 7.18.   *If U is any operator with properties (a) and (b) of Theorem 7.17, U is a* contraction *with modulus β, that is, $\|Uu - Uv\| \leqslant \beta\|u - v\|$, so that, from the fixed-point theorem, U has a unique* fixed point *u^*, and*

$$\|U^n u - u^*\| \leqslant \beta^n \|h - u^*\|$$

for all n (see Theorem 8.1).

PROOF. $v \leqslant u + \|u - v\|$ yields

$$Uv \leqslant U(u + \|u - v\|) = Uu + \beta\|u - v\| \tag{7.15}$$

using (a) and (b) of Theorem 7.17. Interchanging u and v, we obtain $Uu \leqslant Uv + \beta\|u - v\|$. ∎

We shall now state our principal results on optimal plans:

THEOREM 7.19.   (a) *For any Markov $\pi = (f_1, f_2, \ldots)$, denoting by $L(f_n)$ the operator associated with f_n and by $U = \sup L(f_n)$ the operator associated with π, the fixed point u^* of U is the optimal return among π-generated plans: $V(\pi') \leqslant u^*$ for every π-generated π'; and for every $\varepsilon > 0$ there is a π-generated f such that*

$V(f^{\infty}) \geqslant u^* - \varepsilon$. *Any* f *with* $L(f)u^* \geqslant u^* - \varepsilon(1 - \beta)$ *satisfies this inequality*

(b) For any $p \in P(S)$, there is a (p, ε)-optimal plan that is stationary.

(c) *For any* $\varepsilon \geqslant 0$, *if there is an* ε-*optimal* $\pi^* = (\pi_1, \pi_2, \ldots)$, *there is an* $\varepsilon/(1 - \beta)$-*optimal plan that is stationary.*

(d) *Denote for each* $a \in A$ *by* $L(a)$ *the operator associated with* $f \equiv a$: *Any* u *with* $L(a)u \leqslant u$ *for all* a *is an upper bound on returns:* $V(\pi) \leqslant u$ *for all* π.

(e) *If for every* $\varepsilon > 0$, *there is an* ε-*optimal plan, then the optimal return* u^* *is a Baire function and satisfies the* optimality equation $u^* = \sup_a L(a)u^*$.

(f) *A* π *is optimal if and only if its return* $V(\pi)$ *satisfies the* optimality equation.

PROOF. *Proof of* (a). For any π-generated $\pi' = (g_1, g_2, \ldots)$, we have $V(\pi') = L(g_1) \cdots L(g_n)u_n$, where $u_n = V(g_{n+1}, g_{n+2}, \ldots)$. Since each $L(g_i)$ is a contraction with modulus β,

$$\|L(g_1) \cdots L(g_n)u_n - L(g_1) \cdots L(g_n)u^*\|$$
$$\leqslant \beta^n \|u_n - u^*\| \leqslant \beta^n([\|r\|/(1 - \beta)] + \|u^*\|)$$

Thus $L(g_1) \cdots L(g_n)u^* \to V(\pi')$ as $n \to \infty$. But $L(g_1) \cdots L(g_n)u^* \leqslant U^n u^* = u^*$, so that $V(\pi') \leqslant u^*$. From part (d) of Theorem 7.17, there is a π-generated f for which $L(f)u^* \geqslant U u^* - \varepsilon' = u^* - \varepsilon'$, where $\varepsilon' = \varepsilon(1 - \beta)$. We verify inductively that

$$L^n(f)u^* \geqslant u^* - \varepsilon'(1 + \beta + \cdots + \beta^{n-1}) \tag{7.16}$$

for all $n \geqslant 1$. Since $L^n(f)u^* \to V(f^{\infty})$, we conclude that

$$V(f^{\infty}) \geqslant u^* - [\varepsilon'/(1 - \beta)] = u^* - \varepsilon \tag{7.17}$$

Proof of (b). From Corollary 7.13, there is a $(p, \varepsilon/2)$-optimal Markov $\pi = (f_1, f_2, \ldots)$. From (a), there is a stationary f^{∞} with

$$V(f^{\infty}) \geqslant u^* - \varepsilon/2 \geqslant V(\pi^*) - \varepsilon/2,$$

where u^* is the fixed point of the U associated with π. This f^{∞} is (p, ε)-optimal.

Proof of (c). For any $\pi^* = (\pi_1, \pi_2, \ldots)$, $V(\pi^*) = \pi_1 q(r + \beta w)$, where $w \in M(SAS)$, $w(s, a, s') = V(\pi_{s,a})(s')$ and $\pi_{s,a}$ denotes the plan that π^* specifies, starting with the second time, when the first state and act are s, a, that is, $\pi_{s,a} = (\pi_1', \pi_2', \ldots)$, where

$$\pi_n'(\, |s_1, a_1, \ldots, s_n) = \pi_{n+1}(\, |s, a, s_1, a_1, \ldots, s_n) \tag{7.18}$$

If π^* is ε-optimal, $w(s, a, s') \leqslant V(\pi^*)(s') + \varepsilon$ for all s', so that

$$V(\pi^*) \leqslant \pi_1 q(r + \beta V(\pi^*) + \beta\varepsilon) = \pi_1 h, \text{ say.}$$

From Lemma 7.6, there is an f for which $fh \geqslant \pi_1 h$ for all s, so that, for $L(f)$, $V(\pi^*) \leqslant L(f)V(\pi^*) + \beta\varepsilon$. By induction on n we obtain

$$L^n(f)V(\pi^*) \geqslant V(\pi^*) - \varepsilon(\beta + \cdots + \beta^n)$$

Letting $n \to \infty$ yields $V(f^\infty) \geqslant V(\pi^*) - \beta\varepsilon/(1 - \beta)$. Since π^* is ε-optimal, f^∞ is $\varepsilon + [\beta\varepsilon/(1 - \beta)] = \varepsilon/(1 - \beta)$-optimal.

Proof of (d). For any $s_0 \in S$ and any $\varepsilon > 0$, there is a stationary f^∞ such that

$$V(\pi)(s_0) \leqslant V(f^\infty)(s_0) + \varepsilon \qquad (7.19)$$

for all π; simply choose an f^∞ that is (p, ε)-optimal, where p is concentrated on s_0. $L(a)u \leqslant u$ for all a implies $L(f)u \leqslant u$ for a particular f. Thus $L^n(f)u$ decreases to $V(f^\infty)$ and $V(f^\infty) \leqslant u$. Then $V(\pi)(s_0) \leqslant u(s_0) + \varepsilon$. Letting $\varepsilon \to 0$ completes the proof.

Proof of (e). From (c), the hypothesis implies that there is a $1/n$-optimal stationary plan, f_n^∞ say. With $\pi = (f_1, f_2, \ldots)$, the fixed point of the U associated with π is, from (a), the optimal return among π-generated strategies. In particular, $u^* \geqslant V(f_n^\infty)$, so that $u^* \geqslant V(\pi)$ for all π, and u^* is the optimal return. We have $\sup_a L(a)u^* \geqslant U u^* = u^*$. On the other hand, for any $a \in A$,

$$L(a)u^* \leqslant L(a)(V(f_n^\infty) + 1/n) = V(a, f^\infty) + \beta/n \leqslant u^* + \beta/n \qquad (7.20)$$

where (a, f^∞) is the Markov strategy $(g, f, f, f, \ldots)$ with $g \equiv a$. Letting $n \to \infty$ yields $L(a)u^* \leqslant u^*$. Thus u^* satisfies the optimality equation.

Proof of (f). If $V(\pi^*)$ satisfies the optimality equation, we find from (d), with $u = V(\pi^*)$, that π^* is optimal. Conversely, if π^* is optimal, the hypothesis of (e) is satisfied, so that u^*, the optimal return, satisfies the optimality equation. ∎

We note that part (d) of Theorem 7.19 is extremely useful is proving optimality. That is, if u is known to be the return from a strategy π, and u satisfies $L(a)u \leqslant u$ for all a, then (d) implies that π is optimal.

7.4. Special Cases

In this section we introduce the concepts of essential countability and essential finiteness, from which we secure strong results. Furthermore, we establish two improvement routines of Howard and Eaton–Zadeh for the case of finite S, A.

If A is countable, with element $a_1, a_2, \ldots$, every Markov plan is π^*-generated, where $\pi^* = (g_1, g_2, \ldots)$ and $g_n \equiv a_n$. Conversely, for any pure Markov $\pi = (f_1, f_2, \ldots)$, the study of π-generated plans can be reduced to the countable A case by interpreting act n in state s as the selection of $f_n(s)$. We prefer to keep the original A and introduce the concept of essential countability as follows.

DEFINITION 7.20.   *Two acts a and ℓ are equivalent at state s if*
$$r(s, a,) = r(s, \ell,) \quad and \quad q(\, |s, a) = q(\, |s, \ell),$$
that is, if $L(a)u(s) = L(\ell)u(s)$ for all $u \in M(S)$.

DEFINITION 7.21.   *For any Markov $\pi = (f_1, f_2, \ldots)$, A is essentially countable by π if, for every (s, a), there is an n for which $f_n(s)$ is equivalent to a at s.*

DEFINITION 7.22.   *For any Markov $\pi = (f_1, f_2, \ldots)$, A is essentially finite by π if S is partitioned into Borel sets $S_1, S_2, \ldots$, such that, for every (s, a) with $s \in S_n$, at least one of the acts $f_1(s), \ldots, f_n(s)$ is equivalent to a at s.*

THEOREM 7.23.   (a) *If A is essentially countable by $\pi = (f_1, f_2, \ldots)$, the fixed point u^* of the operator U associated with π is the optimal return. U is identical with the operator $\sup_a L(a)$, so that u^* is the unique (bounded) solution of the optimality equation. For every $\varepsilon > 0$, there is an ε-optimal stationary plan.*
(b) *If A is essentially finite by $\pi = (f_1, f_2, \ldots)$, there is an optimal stationary plan.*

PROOF.   *Proof of (a).* For any $u, s, L(f_n)u(s) = L(a)u(s)$, where $a = f_n(s)$. Thus $U u \leqslant \sup L(a)u$. But for any $a \in A$, $L(a)u(s) = L(f_n)u(s)$ for some n, so that $L(a)u(s) \leqslant U u(s)$, and $\sup L(a)u(s) \leqslant U u(s)$. Thus the operators $\sup_a L(a)$ and U are identical. Part (d) of Theorem 7.19 then implies that $V(\pi^*) \leqslant u^*$ for all π. From part (a) of Theorem 7.19, there is a stationary f^∞ with $V(f^\infty) \geqslant u^* - \varepsilon$. This f^∞ is ε-optimal.

Proof of (b). If A is essentially finite, define B_n as the set of all s for which n is the smallest i with $L(f_i)u^*(s) = \sup_n L(f_n)u^*(s)$. (The sequence $\{L(f_n)u^*(s)\}$ contains only finitely many different numbers.) Define $f = f_n$ on B_n, so that $L(f)u^* = U u^* = u^*$. Then u^*, as the fixed point of $L(f)$, is the return from f^∞, and f^∞ is optimal. ∎

The case of *finite* S, A was treated in Chapter 2 as Markovian decision processes with discounting. We have also given an improvement routine of Howard (Theorem 2.4). There is also an improvement due to Eaton and Zadeh [53], which we now state.

THEOREM 7.24.   (a) *(Howard improvement.)* If $V(g, \pi) \geqslant V(\pi)$, then $V(g^\infty) \geqslant V(\pi)$.
(b) *(Eaton–Zadeh improvement.)* For any f, g mapping S into A, define $h = f$ on $V(f^\infty) \geqslant V(g^\infty)$, $h = g$ on $V(g^\infty) \geqslant V(f^\infty)$. Then
$$V(h^\infty) \geqslant \max (V(f^\infty), V(g^\infty))$$

PROOF.   *Proof of (a).*   We have $L(g)V(\pi) = V(g, \pi) \geqslant V(\pi)$, so that

$$L_n(g)V(\pi) \to V(g^\infty) \qquad \text{and} \qquad V(g^\infty) \geqslant V(g, \pi)$$

The proof of (a) has been given in Theorem 2.4.

Proof of (b).   We have, for any u,

$$L(h) = L(f)u \qquad \text{on} \quad V(f^\infty) \geqslant V(g^\infty)$$
$$L(h) = L(g)u \qquad \text{on} \quad V(g^\infty) \geqslant V(f^\infty)$$

With $u = \max (V(f^\infty), V(g^\infty))$, we obtain

$$L(h)u = L(f)u \geqslant L(f)V(f^\infty) = V(f^\infty) = u \qquad \text{on} \quad V(f^\infty) \geqslant V(g^\infty)$$
$$L(h)u = L(g)u \geqslant L(g)V(g^\infty) = V(g^\infty) = u \qquad \text{on} \quad V(g^\infty) \geqslant V(f^\infty)$$

Thus $L(h)u \geqslant u$, so that $V(h^\infty) \geqslant u$. ∎

References and Comments

The results of this chapter have been given by Blackwell [16]. The basic facts on probability theory are found in Loéve [79]. Lemma 7.6 has been given by Blackwell [15]. The principal general results on optimal plans of Theorem 7.19 are related to those of Dubins and Savage [51]; Theorem 7.19(c) corresponds to Theorem 3.9.6 of [51], Theorem 7.19(d) to Theorem 2.12.1, 3.3.1, and Theorem 7.19(e) to Theorem 3.3.1.

Further problems have been studied by Strauch [109] and Blackwell [17], where the return is either negative or positive, and the strengthened results are obtained.

De Leve [29, 30] has studied another generalized Markovian decision process and given the corresponding policy iteration algorithm.

Chapter 8

The Principle of Contraction Mappings in Markovian Decision Processes

8.1. Introduction

We have studied some models, and have seen the existence of some common structures in these models. In this chapter we treat the general decision process models, or the dynamic programming models, and discuss contraction and monotonicity properties. In our decision models, if the contraction and monotonicity properties are satisfied, the "fixed point" exists; this fixed point becomes our objective function. Furthermore, if the N-stage contraction, which is a weakened form of the contraction property, and monotonicity properties are satisfied, we also have the same result.

It is sufficient for the later discussion to state briefly some known results concerning a metric space and the principle of contraction mappings. These results, given below, can be found in many standard texts on mathematical analysis.

Consider a set V. A function ρ that maps $V \times V$ to the reals is called a *metric* if

(a) $\rho(u, v) \geqslant 0$ for all $u, v \in V$
(b) $\rho(u, v) = 0$ if and only if $u = v$
(c) $\rho(u, v) \leqslant \rho(u, w) + \rho(w, v)$ for $u, v, w \in V$ (triangle inequality)

Then V is called a *metric space* if ρ is a metric for V. Let A be a map having domain and range both V. The function A is called a contraction mapping if there exists some c $(0 \leqslant c < 1)$ such that $\rho(Au, Av) \leqslant c\rho(u, v)$ for any $u, v \in V$. The element of $v^* \in V$ is called a *fixed point* of A if $Av^* = v^*$. A sequence $\{v_n\}$, $n = 1, 2, \ldots$, the elements of V, is called a *Cauchy sequence* if for every $\varepsilon > 0$ there exists an M such that $\rho(v_m, v_n) < \varepsilon$ for every $m, n > M$. A metric space is said to be *complete* if for every Cauchy sequence $\{v_n\}$, there exists an element $v \in V$ such that $\lim_{n \to \infty} \rho(v_n, v) = 0$. With the above preparation we come to some well-known theorems on mathematical analysis.

THEOREM 8.1 (*the principle of contraction mappings*). *Let V be a complete metric space. Suppose that A is a contraction mapping. The A has a unique fixed point $v^* \in V$, that is, the equation*

$$Av = v \tag{8.1}$$

has a unique solution $v^* \in V$.

A map A^n is defined recursively by $A^1 = A$ and $A^n = A(A^{n-1})$ $(n > 0)$. An extension of this theorem is:

THEOREM 8.2. *Let V be a complete metric space. Suppose that A^N is a contraction mapping for some N. Then, the equation*

$$Av = v \qquad\qquad (8.2)$$

has a unique solution $v^* \in V$.

Applying the principle of contraction mappings, we can find or approximate an optimal strategy in our decision models. That is, the problems that we have treated (not all) in the preceding chapters are those of finding a fixed point; this will be discussed in Section 8.5 with examples.

In Section 8.2 we shall introduce the notation and assumptions for the general decision process and shall show that the contraction and monotonicity assumptions imply the existence of the fixed point. Furthermore, the N-stage contraction and monotonicity assumptions imply the same results. In Section 8.3, using the results of Section 8.2, we shall establish the optimization schemes for finding or approximating an optimal strategy, for example, the successive approximation, mathematical programming, and Howard-type policy iteration. In Section 8.4, two decision problems that are mutually relative will be discussed from the viewpoint of fixed points. In the final section 8.5, we shall present some examples that have been discussed in the preceding chapters.

8.2. Contraction and Monotonicity Assumptions

Consider a nonempty set Ω. An element of Ω is called a point and is denoted by x. Associated with each $x \in \Omega$ is a *decision set* D_x. An element of D_x is called a decision and is denoted by d_x. The *policy space* Δ is defined as the Cartesian product of the decision sets, that is, $\Delta = \underset{x \in \Omega}{\times} Dx$. An element of Δ is called a *policy* and is denoted by δ. Then, a policy can be interpreted as a decision procedure that specifies a decision for each point. Further, any such combination of decisions constitutes a policy.

In order to introduce the return function, let V be the collection of all bounded functions from Ω to the reals, that is, $v \in V$ if and only if v, a map Ω to the reals, and $\underset{x \in \Omega}{\sup} |v(x)| < \infty$. A metric ρ on V is defined by

$$\rho(u, v) = \underset{x \in \Omega}{\sup} |u(x) - v(x)|$$

The space V is complete in this metric.

Let h, the return, be a function giving a real number to each triplet (x, d_x, v) with $x \in \Omega$, $d_x \in D_x$ and $v \in V$. The return $h(x, d_x, v)$ may be considered as the total return for "starting" at point x and choosing d_x with the prospect of receiving $v(z)$ if the pair (x, d_x) causes a "transition" to point z. [Whether $v(z)$ could be realized by any policy is immaterial; $h(x, d_x,)$ describes what the pair (x, d_x) yields as a function of v.] The important contraction assumption is as follows:

Contraction assumption.   For some c satisfying $0 \leqslant c < 1$,

$$|h(x, d_x, u) - h(x, d_x, v)| \leqslant c\rho(u, v) \tag{8.3}$$

for each $u \in V$, $v \in V$, $x \in \Omega$, and $d_x \in D_x$.

Since a contraction mapping is implicit in the contraction assumption, we introduce a function H_δ having the domain V and a range assumed to be contained in V. Let δ_x denote the decision in δ that applies to point x. Then H_δ is defined by

$$[H_\delta(v)](x) = h(x, \delta_x, v) \tag{8.4}$$

where $H_\delta(v)$ is the element of V which H_δ assigns v, and $[H_\delta(v)](x)$ is the real number which the function $H_\delta(v)$ associates with the point x. The contraction assumption is equivalent to the following: For some c satisfying $0 \leqslant c < 1$,

$$\rho[H_\delta u, H_\delta v] \leqslant c\rho[u, v] \tag{8.5}$$

for each $u, v \in V$ and $\delta \in \Delta$. Hence, H_δ is a contraction mapping, and Theorem 8.1 guarantees that H_δ has a unique fixed point v_δ. That is, for each policy δ, there exists a unique element $v_\delta \in V$ such that

$$v_\delta(x) = h(x, \delta_x, v_\delta) \tag{8.6}$$

for each $x \in \Omega$.

The function v_δ is called the return function of the policy $\delta \in \Delta$. The *optimal return function* is defined by

$$f(x) = \sup_{\delta \in \Delta} v_\delta(x) \tag{8.7}$$

The following simply inequality will be useful.

THEOREM 8.3.   *Suppose that the contraction assumption is satisfied. For any $\delta \in \Delta$ and any $v \in V$, we have*

$$\rho(v_\delta, v) \leqslant \rho(H_\delta v, v)/(1 - c) \tag{8.8}$$

PROOF.   Applying the triangle inequality, we have

$$\rho(H_\delta{}^n v, v) \leqslant \sum_{i=1}^{n} \rho(H_\delta{}^i v, H_\delta{}^{i-1} v)$$

$$\leqslant \sum_{i=1}^{n} c^{i-1} \rho(H_\delta v, v) \leqslant \rho(H_\delta v, v)/(1 - c) \tag{8.9}$$

Since $\rho(v_\delta, v) \leqslant \rho(v_\delta, H_\delta{}^n v) + \rho(H_\delta{}^n v, v)$, and $\rho(v_\delta, H_\delta{}^n v) \to 0$, we have $\rho(v_\delta, v) \leqslant \rho(H_\delta v, v)/(1 - c)$. ∎

In the above discussion the concept of maximization was not considered. Here, a map A having domain V is defined by

$$(Av)(x) = \sup_{d_x \in D_x} h(x, d_x, v) \tag{8.10}$$

for each $v \in V$ and $x \in \Omega$. We also assume that the range of A is contained in V. The next theorem states that A is a contraction mapping.

THEOREM 8.4.   *Suppose that the contraction assumption is satisfied. For each $u \in V$ and $v \in V$, we have*

$$\rho(Au, Av) \leqslant c\rho(u, v) \tag{8.11}$$

PROOF.   Consider arbitrary u, v, and x, and write $(Au)(x) = (Av)(x) + k$. Consider the case $k > 0$. For each integer n, let $d_x{}^n$ be an element of D_n such that $h(x, d_x{}^n, u) \geqslant (Au)(x) - k/n$. Clearly,

$$(Au)(x) - k/n \geqslant (Av)(x) \geqslant h(x, d_x{}^n, v)$$

the last by definition. Combining inequalities yields

$$0 \leqslant (Au)(x) - (Av)(x) - k/n \leqslant h(x, d_x{}^n, u) - h(x, d_x{}^n, v) \leqslant c\rho(u, v) \tag{8.12}$$

Since the preceding is true for each n, $|(Au)(x) - (Av)(x)| \leqslant c\rho(u, v)$; this inequality is trivial for $k = 0$ and is similarly established for $k < 0$. ∎

Theorem 8.1 guarantees that A has a unique fixed point, that is, that there exists exactly one element $v^* \in V$ such that

$$v^*(x) = \sup_{d_x \in D_x} h(x, d_x, v^*) \tag{8.13}$$

for each $x \in \Omega$.

Equation (8.13) is, in rather general notation, a "functional equation" of dynamic programming (see Bellman [7]). We have not yet seen whether $v^* = f$. After introducing the monotonicity assumption, we shall show that $v^* = f$. The choice of maximization in defining A is arbitrary; Theorem 8.4 holds with $(Av)(x) = \inf_{d_x \in D_x} h(x, d_x, v)$.

COROLLARY 8.5.   *For $\varepsilon > 0$, there exists a policy δ such that*

$$\rho[H_\delta(v^*), v^*] \leqslant \varepsilon(1 - c)$$

and any such δ satisfies $\rho(v_\delta, v^*) \leqslant \varepsilon$. If $\rho[H_\delta(v^*), v^*] = 0$, then $v_\delta = v^*$.

PROOF.   For $\varepsilon > 0$, the existence of a policy δ such that

$$\rho[H_\delta(v^*), v^*] \leqslant \varepsilon(1 - c)$$

follows directly from (8.13). Substituting for v in Theorem 8.3 yields $\rho(v_\delta, v^*) \leqslant \rho[H_\delta(v^*), v^*]/(1 - c) \leqslant \varepsilon$ for $\varepsilon \geqslant 0$. ∎

Corollary 8.6 is established by noting that extrema of continuous functions over compact sets are attained and then by applying the last part of Corollary 8.5. We note that this approach to Corollary 8.6 circumvents the usual recourse to Tychnoff's theorem.

COROLLARY 8.6. *Suppose that, for each fixed x, $h(x, \,, v^*)$ is a continuous function of d_x in a topology for which D_x is compact. Then there exists a policy such that $v_\delta = v^*$.*

With B as an operator on V, define the modulus of B as the smallest number c such that $\rho(Bu, Bv) \leqslant c\rho(u, v)$ for each $u, v \in V$.

With I as an arbitrary nonempty set, suppose that $\{B_\alpha : \alpha \in I\}$ is a collection of operators on V, each of which has modulus c or less, define the function E having domain and range contained in V by $(Ev)(x) = \sup_{\alpha \in I} (B_\alpha v)(x)$. Then an argument similar to that for Theorem 8.4 establishes that E has modulus c or less.

We next assume that for $u, v \in V$, we write $u \geqslant v$ if $u(x) \geqslant v(x)$ for each $x \in \Omega$, and $u > v$ if $u \geqslant v$ and $u \neq v$, which refers to the definition of vector inequality in Section 2.2.

Monotonicity assumption. If $u \geqslant v$, then $H_\delta(u) \geqslant H_\delta(v)$ for each $\delta \in \Delta$.

The assumption is equivalent to $h(x, d_x, u) \geqslant h(x, d_x, v)$ if $u \geqslant v$. The next theorem guarantees that $v^* = f$.

THEOREM 8.7. *Suppose the monotonicity and contraction assumptions are satisfied. Then $v^* = f$.*

PROOF. From Corollary 8.5 we know that $v^* \leqslant f$. Since $H_\delta v^* \leqslant v^*$ for each δ, recursive application of the monotonicity assumption yields $H_\delta{}^n v^* \leqslant v^*$ for each δ. Since $f(x) = \sup_\delta v_\delta(x)$, this implies $f \leqslant v^*$. ∎

Theorem 8.7 states that the solution of (8.13) is unique and is the optimal return function. A policy δ is called ε-*optimal* if $\rho(v_\delta, f) \leqslant \varepsilon$ and *optimal* if $v_\delta = f$. Corollary 8.5 and Theorem 8.7 demonstrate the existence of an ε-optimal policy and Corollary 8.6 gives sufficient conditions for the existence of an optimal policy.

If the sequence $\{v_n\}$, $n = 0, 1, \ldots$, satisfies $v_n \geqslant v_{n-1}$ for each n, we write

$\{v_n\} \uparrow$. Lemma 8.8 exhibits a useful consequence of the monotonicity assumption.

LEMMA 8.8. *Suppose that the monotonicity assumption is satisfied. If $u \geqslant v$, then $Au \geqslant Av$. If $Av \geqslant v$, then $\{A^n v\} \uparrow$. If $H_\delta v \geqslant v$, then $\{H_\delta^n v\} \uparrow$.*

PROOF. By definition, $Au \geqslant H_\delta u$. Suppose that $u \geqslant v$. Then, $H_\delta u \geqslant H_\delta v$ implying $Au \geqslant H_\delta v$ for each δ; hence, $Au \geqslant Av$. If $Av \geqslant v$, then recursive application of the preceding statement yields $\{A^n v\} \uparrow$. If $H_\delta v \geqslant v$, then recursive application of the monotonicity assumption yields $\{H_\delta^n v\} \uparrow$. ∎

Although we have seen that the terminating process in Section 3.7 does not satisfy the contraction assumption (these facts will be stated in Section 8.5), it has the same structure as the discounted model that satisfies the contraction assumption. Here we shall develop the N-stage contraction assumption, which is the weakened form of the contraction assumption.

N-stage contraction assumption. For each δ the operator H_δ^N has modulus c or less, where c and N are independent of δ and $c < 1$. Furthermore, for each δ, H_δ has modulus 1 or less.

The terminating process satisfies the N-stage contraction assumption, which will be given in Section 8.5.

As before, we assume that the range of A is contained in V. Define v_δ as a unique fixed point of the contraction mapping H_δ^N; it follows that v_δ is a unique fixed point of H_δ. Since H_δ has modulus 1 or less, the triangle inequality implies $\rho(H_\delta^N v, v) \leqslant N\rho(H_\delta v, v)$. Hence, as in Theorem 8.3, $\rho(v_\delta, v) \leqslant \rho(H_\delta v, v)N/(1 - c)$.

Next, define f by $f(x) = \sup_\delta v_\delta(x)$. To show that $f \in V$, define the function E having domain V by $(Ev)(x) = \sup_\delta (H_\delta^N v)(x)$. Since $H_\delta^N v \leqslant Ev \leqslant A^N v$, the last by the monotonicity assumption, E has range contained in V. Then, the N-stage contraction assumption and observation at the statement of the preceding suffice for E to be a contraction mapping. Let v^* be a unique fixed point of E. Since $H_\delta^N v^* \leqslant v^*$, we have $v_\delta \leqslant v^*$ for each δ, implying $f \leqslant v^*$. Hence $f \in V$. Parts (a)–(c) of the following theorem have just been established. Proofs of (d) and (e) are postponed until after Lemma 8.10.

THEOREM 8.9. *Suppose that the monotonicity and N-stage contraction assumption are satisfied. Then:* (a) v_δ *is a unique fixed point of* H_δ. (b)

$\rho(v_\delta, v) \leqslant \rho(H_\delta v, v)N/(1 - c)$. (c) E *is a contraction mapping of modulus* c *or less.* (d) f *is a unique fixed point of E and of A.* (e) *If* $v \leqslant f$, *then*

$$\rho(A^N v, f) \leqslant c\rho(v, f)$$

Lemma 8.10 will be useful both for Theorem 8.9 and for the optimization schemes in the next section. We shall prove only (a), since (b) and (c) are obvious. Let $\vec{1}$ be the unit function from Ω to the reals defined by $\vec{1}(x) = 1$ for each $x \in \Omega$.

LEMMA 8.10. *Suppose that the monotonicity and N-stage contraction assumptions are satisfied. Then: (a) If $Av \leqslant v$, then $v \geqslant f$; if $Av \geqslant v$, then $v \leqslant f$. (b) $Av_\delta \geqslant v_\delta$ for each $\delta \in \Delta$. (c) If $H_\delta v \geqslant v$, then $v_\delta \geqslant H_\delta v$.*

PROOF. [Lemma 8.10(a).] First, suppose that $Av \leqslant v$. Then, $H_\delta v \leqslant v$ for each δ, which implies that $H_\delta^n v \leqslant v$ for each n. Hence, $v_\delta \leqslant v$ for each δ, which implies that $v \geqslant f$.

Next, since $Af \geqslant H_\delta f \geqslant H_\delta v_\delta = v_\delta$ for each δ, we have $Af \geqslant f$. Suppose that $Av \geqslant v$. Define u by $u(x) = \max\{v(x), f(x)\}$. Then $Au \geqslant Av$ and $Au \geqslant Af$, which implies that $Au \geqslant u$. For arbitrary positive ε, pick δ such that $H_\delta u \geqslant Au - \varepsilon\vec{1}$. We wish to prove that $H_\delta^n(Au) \geqslant Au - n\varepsilon\vec{1}$. Since $Au \geqslant u$, this holds for $n = 1$. Suppose that it holds for n. Then

$$H_\delta^{n+1}(Au) \geqslant H_\delta(Au - n\varepsilon\vec{1}) \geqslant H_\delta(Au) - n\varepsilon\vec{1} \geqslant Au - (n + 1)\varepsilon\vec{1}$$

which proves the relation by induction. Since $v_\delta \leqslant f \leqslant u \leqslant Au$, the N-stage contraction assumption assures $H_\delta^N(Au) \leqslant v_\delta + c\rho(Au, v_\delta)\vec{1}$. Combining inequalities, $0 \leqslant Au - v_\delta \leqslant [N\varepsilon + c\rho(Au, v_\delta)]\vec{1}$. Suppose that $Au > f$. Then take $\varepsilon = \rho(Au, v_\delta)(1 - c)/2N$. By substitution,

$$\rho(Au, v_\delta) \leqslant \rho(Au, v_\delta)(1 + c)/2 < \rho(Au, v_\delta)$$

which is a contradiction. Hence, $Au = f$ and $v \leqslant f$. ∎

PROOF. [Theorem 8.9.] Parts (a)–(c) are established. As noted previously, $Af \geqslant f$. Then $A(Af) \geqslant Af$ and the first part of Lemma 8.10 implies that $Af \leqslant f$. Hence, $Af = f$. If $Ag = g$, then Lemma 8.10 implies that $g = f$; hence f is a unique fixed point of A.

We have established $f \leqslant v^*$. To show $v^* = f$, define u by

$$u(x) = \max_{0 \leqslant n < N} (A^n v^*)(x)$$

Then, $Au \geqslant A^n v^*$ for $1 \leqslant n \leqslant N$. Since $A^N v^* \geqslant Ev^* = v^*$, we have $Au \geqslant u$.

9

Hence, by Lemma 8.10, $u \leqslant f$, which implies that $v^* \leqslant f$. Hence $v^* = f$. For (e), note that $A^N v \geqslant Ev$. If $v \leqslant f$, then $Ev \leqslant A^N v \leqslant f = Ef$, which implies that $\rho(A^N v, f) \leqslant \rho(Ev, Ef) \leqslant c\rho(u, v)$. ∎

8.3. Optimization Schemes

In this section we develop three techniques for determining or approximating v^* and for finding policies whose returns approximate or attain v^*. One technique is an application of "successive approximations" of mathematical analysis; the second technique provides general mathematical programming equivalents; and the third technique generalizes one of the Howard-type policy iteration algorithms.

The first technique uses the contraction assumption, but not the monotonicity one. Suppose that the contraction property is satisfied. Theorem 8.4 implies that $\rho(A^n v, v^*) \to 0$ for any v; this means that v^* can be approximated by successive applications of A to any initial vector v_0. We can compute a bound of optimal policies from the preceding results, but we omit it here.

The second technique gives the *mathematical programming* formulations. If the monotonicity and the N-stage contraction assumptions are both satisfied, mathematical programming formulations are available. Let "min v" denote the function whose value at x is the smallest value of $v(x)$ over those v's satisfying the constraint. Consider the following mathematical programming problem:

$$\min v \tag{8.14}$$

subject to

$$Av \leqslant v \tag{8.15}$$

Since $Af = f$, f is feasible for the above problem. By part (a) of Lemma 8.10, f is optimal for the above problem. It is clear that, if Ω has finitely many points, min v is equivalent to minimizing $\sum_{x \in \Omega} v(x)$ or any other positive combination of the $v(x)$. The above problem can be written equivalently as "min v subject to $h(x, d_x, v) \leqslant v(x)$ for each x and d_x," which has been given in the preceding chapters as the dual problems.

The third technique is a generalization of the Howard-type *policy iteration algorithms*. Suppose that the monotonicity and N-stage assumptions are satisfied and that Av is attained for each v; that is, $Av = H_\gamma v$ for some policy γ that may depend on v. The "n" in the policy improvement routine given may be any positive integer.

1. Pick any initial δ.

2. Calculate v_δ.

3. Calculate $u = A^{n-1}v_\delta$; then $v = Au = H_\gamma u$.

4. If $\rho(u, v) > \varepsilon$, replace δ by γ and go to Step 2. If $\rho(u, v) \leqslant \varepsilon$, calculate v_δ and stop.

Lemmas 8.8 and 8.10 imply $v_\delta \leqslant A^{n-1}v_\delta = u \leqslant Au = H_\gamma u \leqslant v_\gamma$.

8.4. Symmetries

As we have seen in the preceding discussion, we have two similar problems, for example, the discounted Markovian decision problem and the discounted semi-Markovian decision problem. The former is a special case of the latter. We have also seen that two optimization algorithms for these problems are very similar. In this section we consider such problems.

Consider two optimization problems—problem U and problem P—of the type we have been discussing. Problem U is described using unprimed notation—for example, Ω, D_x, h, etc., and problem P is described using primed notation—for example, Ω', D_x', h', etc. Suppose $\Omega = \bigcup_{x \in \Omega} E_x$, where $\{E_x\}_{x \in \Omega}$ is a collection of nonempty pairwise disjoint subsets of Ω'. Roughly speaking, E_x is the subset of Ω' containing those points which are "equivalent" to x; it may correspond to the various histories of x. Let e be the map of V into V' defined by $[e(v)](z) = v(x)$ for each $z \in E_x$ and x. Roughly speaking, the function e maps v into that function whose value is $v(x)$ at each point which is equivalent to x. Problem P is said to be *generated* from problem U if, in addition to the above, (i) $D_z' = D_x$ for each $z \in E_x$ and each x, and (ii) $h'[z, d_x, e(v)] = h(x, d_x, v)$ for each $z \in E_x$, each x, d_x, and v. It is convenient to introduce the map w of Δ into Δ' defined by $[w(\delta)]_z = \delta_x$ for each $z \in E_x$ each x and each δ. The proof of Theorem 8.11 is routine and is omitted.

THEOREM 8.11. *Let problem P be generated from problem U and suppose that problem P satisfies either (a) the contraction assumption or (b) both the N-stage contraction and the monotonicity assumptions. Then, problem U satisfies the same assumption(s) and has a unique fixed point v^*. Furthermore, $e(v^*)$ is a unique fixed point of problem P, and $\rho[v_{w(\delta)}, e(v^*)] = \rho(v_\delta, v^*)$ for each $\delta \in \Delta$.*

In other words, the fixed points for problems U and P are essentially the same, and a policy δ whose return approximates or attains v^* has an equivalent policy $w(\delta)$ whose return approximates or attains $e(v^*)$. Thus, for the purpose of determining fixed points and ε-optimal policies, we can study problem U rather than the more complex problem P.

8.5. Examples

In this section we shall consider the models discussed in the preceding chapters as examples of general decision models. The models are the discounted Markovian decision model, the terminating model, the generalized Markovian decision model, the discounted semi-Markovian decision model and the terminating stochastic game. The first four models have been discussed in the preceding chapters and the stochastic game will be discussed in the Appendix.

EXAMPLE 1. (Chapter 2: Discounted Markovian decision model.) For this model, we have $\Omega = S = \{1, 2, \ldots, N\}$ and $D_i = K_i = \{1, 2, \ldots, K_i\}$ for each $i \in S$. A *policy* refers to a function f whose ith element $f(i)$ is the decision in state i, where $f(i) \in K_i$. That is, $\delta \in \Delta$ corresponds to $f \in F$. Note that a *strategy* defined in Chapter 2 corresponds to a policy defined in this chapter. Further we have used in Chapter 5 a policy which is a sequence of decisions backward in time. In this chapter we shall use conveniently a policy which is indeed a strategy used in the preceding chapters. Then

$$h(i, f, v) = r_i^k + \beta \sum_{j=1}^{N} p_{ij}^k v_j$$

where β $(0 \leqslant \beta < 1)$ is a discount factor and $k = f(i)$.

First, we have

$$|h(i, f, u) - h(i, f, v)| = \beta \left| \sum_{j=1}^{N} p_{ij}^k (u_j - v_j) \right|$$

$$\leqslant \beta \sum_{j=1}^{N} p_{ij}^k |u_j - v_j| \leqslant \beta \rho(u, v)$$

which shows that the contraction assumption is satisfied.

Second, we have

$$h(i, f, u) - h(i, f, v) = \beta \sum_{j=1}^{N} p_{ij}^k (u_j - v_j) \geqslant 0$$

if $u \geqslant v$, which shows that the monotonicity assumption is satisfied.

Thus, applying the results of Section 8.3, we have two optimization schemes for finding an optimal strategy. One is a linear programming algorithm that was given in Section 2.3 as the dual problem. Another is the policy iteration algorithm that was given in Section 2.2.

EXAMPLE 2 (Section 3.7: Terminating process). For this model, we have $\Omega = S = \{2, 3, \ldots, N\}$ and $D_i = K_i = \{1, 2, \ldots, K_i\}$ for each $i \in S$. A policy is defined by f whose ith element $f(i)$ is the decision in state $i \in S$. Then

$$[H_f(v)](i) = h(i, f, v) = r_i^k + \sum_{j=2}^{N} p_{ij}^k v_j$$

The model does not satisfy the contraction assumption, but H_f^N satisfies the contraction assumption, that is, the model satisfies the N-stage contraction assumption from the terminating assumption. We note that the "N" in H_f^N is equal to the number of states. Of course, the model also satisfies the monotonicity assumption. Thus we can apply Theorem 8.9 and the optimization schemes are available similar to that which has been given in Section 3.7.

EXAMPLE 3. (Chapter 7: Generalized Markovian decision process.) For this model, the state space is a nonempty Borel set $\Omega = S$ and the policy space is a nonempty Borel set $\Delta = A$, and their elements are denoted by $s \in S$ and $a \in A$, respectively. Let qr be the expectation of the immediate return associated with making decision a at point s, that is,

$$qr = \int_{s' \in S} r(s, a, s')\, dq(s'|s, a)$$

and let $q(\ |s, a)$ be the probability measure over S determined by the pair (s, a) with $q(S|s, a) = 1$. With a discount factor $\beta\ (0 \leqslant \beta < 1)$ and $K(v) = \{u \in V : u \geqslant v, u \text{ integrable}\}$, let

$$h(s, a, v) = qr + \beta\psi\ (v : s, a)$$

$$\psi(\ v : s, a) = \sup_{u \in K(v)} \int_{s' \in S} u(s')\, dq(s'|s, a)$$

The property of ψ which we shall verify and then exploit is that for each fixed s and a, ψ is subadditive, that is, $\psi(u + v : s, a) \leqslant \psi(u : s, a) + \psi(v : s, a)$ Fixing s and a for the remainder of the discussion, we abbreviate $\psi(u : s, a)$ by the symbol $\psi(u)$. As defined,

$$K(u + v) \supset \{u' + v' : u' \geqslant u, v' \geqslant v, u', v' \quad \text{integrable}\}$$

Hence, by set inclusion,

$$\psi(u + v) \leqslant \sup_{u' \in K(u),\, v' \in K(v)} \int_{s' \in S} [u'(s') + v'(s')]\, dq(s'|s, a) = \psi(u) + \psi(v)$$

Then, $\psi[(u - v) + v] \leqslant \psi(u - v) + \psi(v)$, or $\psi(u) - \psi(v) \leqslant \psi(u - v)$. As ψ is defined, it may be that $\psi(u) \neq -\psi(-u)$. However, $|\psi(u)| \leqslant \sup_{s \in S} u(s)$. Then

$$|\psi(u) - \psi(v)| \leqslant \max \{|\psi(u - v)|, \qquad |\psi(v - u)|\} \leqslant \rho(u, v)$$
$$|h(s, a, u) - h(s, a, v)| = \beta|\psi(u) - \psi(v)| \leqslant \beta\rho(u, v)$$

verifying that the contraction assumption is satisfied. The monotonicity assumption can be routinely verified. Note that f might not be measurable and that the same definition of ψ works for the maximizing and minimizing problems.

EXAMPLE 4.   (Section 6.4: Discounted Semi-Markovian decision model.) This example differs from the preceding three in that the process is *continuous in time*. This model is similar to Example 1. The function "ψ" is again used to circumvent some integrability problems, and the model is shown to be generated from Example 1, which has no integrability problems.

Let $\Omega' = \{1, 2, \ldots, N\} \times$ reals and $x' = (i, t)$, with $1 \leqslant i \leqslant N$ and $t \in$ reals. Let $D_i' = D_{(i, t)} = \{1, 2, \ldots, K_i\}$, independent of t. Selection of decision k from D_i' at point (i, t) causes a transition to some state at some time greater than t. It is convenient to represent this phenomenon by two random variables. Let $P_r[X_{i, k} = j] = p_{ij}^k$ be the probability that transition occurs to state j given that decision k is made in state i. Let

$$P_r[X_{i, j, k} < u] = F_{ij}^k(u)$$

be the probability that the interval of time to transition is less than u, given that the decision k is made in state i and given the condition that transition will occur to state j. Both $X_{i, k}$ and $X_{i, j, k}$ are independent of t; the present value at time t (not time 0) of the return from time t on is

$$h'[(i, t), k, v'] = r_i^k + \sum_{j=1}^{N} p_{ij}^k \int_0^\infty v'(j, t + u)\, e^{-\alpha u}\, dF_{ij}^k(u)$$

The above expression is technically correct only if v' is integrable. In general, we repeat "$\int$" by "ψ" and assume $\int dF_{ij}^k(u)\, e^{-\alpha u} \leqslant c < 1$ for each i, j, and k. The contraction assumption is then satisfied, and the monotonicity assumption is again routinely verified.

We can readily check that Example 4 is generated from Example 1 with $\Omega = \{1, 2, \ldots, N\}$, $D_i = \{1, 2, \ldots, K_i\}$, $P_r[j : i, k] = p_{ij}^k \int dF_{ij}^k(u)\, e^{-\alpha u}$, $E_i = \{(i, t) : t \in$ reals$\}$, and r_i^k unchanged. Thus, Theorem 8.11 allows us to investigate the optimization problem in a far simpler environment having a finite number of points and decisions and no integrability difficulties. Then, Corollary 8.6 guarantees the existence of a stationary optimal strategy for this example, which is the proof of Theorem 6.2.

EXAMPLE 5. (Appendix: Terminating stochastic game.) Let $\Omega = \{1, 2, \ldots, N\}$, where point i is thought of as the nth of a collection of N rectangular games. Let r_i^{kr} be the return from player II to player I if they play game i and choose pure strategies k and r, respectively, with $1 \leqslant k \leqslant m_i$, and $1 \leqslant r \leqslant n_i$. Let p_{ij}^{kr} be the probability that the next game played in game i, given that game i is now played, and that pure strategies k and r are chosen by players I and II, respectively. For a terminating stochastic game, we assume

$$\sum_{j=1}^{N} p_{ij}^{kr} = 1 - s_i^{kr}, \qquad s_i^{kr} > 0$$

for each k, r, and i. Let $x_i(y_i)$ be a mixed strategy for player I (II) for game i. With strategies x_i and y_i for game i, and with v as the terminating return function, the one-stage expected return function h is given by

$$h(i, x_i, y_i, v) = \sum_{k=1}^{m_i} \sum_{r=1}^{n_i} x_i^k y_i^r \left\{ r_i^{kr} + \sum_{j=1}^{N} p_{ij}^{kr} v(j) \right\}$$

Note that with fixed i and v, the above is the payoff function of a rectangular game whose element is the term in the brackets, "$\{\ \}$". Hence, by the minimax theorem for rectangular games, an operator B on V is defined by

$$(Bv)(i) = \max_x \min_y h(i, x, y, v) = \min_y \max_x h(i, x, y, v)$$

Let D_i and O_i be the sets of all randomizations over $\{1, 2, \ldots, m_i\}$ and $\{1, 2, \ldots, n_i\}$, respectively. Let $\Delta = \underset{i=1}{\overset{N}{\times}} D_i$ and $\Pi = \underset{i=1}{\overset{N}{\times}} O_i$, with $\delta \in \Delta$ and $\pi \in \Pi$. (Then, δ is an N-tuple of probability distributions.) For each fixed δ and π, $H_{\delta, \pi}$ obeys the contraction assumption (see appendix) and, hence, is a contraction mapping; let $v_{\delta, \pi}$ be its fixed point. Define H_δ by

$$(H_\delta v)(i) = \min_\pi [H_{\delta, \pi}(v)](i)$$

Theorem 8.4, though minimizing instead of maximizing, guarantees that H_δ satisfies the contraction assumption for each δ. Then, noting that

$$(Bv)(i) = \max_\delta (H_\delta v)(i)$$

and reapplying Theorem 8.4 assures that $H_{\delta, \pi}$ satisfies the monotonicity assumption. Then, f is a unique solution of:

(a) $v(i) = \max_x \min_y h(i, x, y, v)$

(b) $v(i) = \min_y \max_x h(i, x, y, v)$

(c) $f = \max_\delta \min_\pi v_{\delta, \pi}$

(d) $f = \min_\pi \max_\delta v_{\delta, \pi}$

References and Comments

The results of this chapter have been given by Denardo [31].† A monotonicity assumption was first introduced by Mitten [91] in 1964. The well-known facts on mathematical analysis are found in, for example, Kolmogorov and Fomin [76], and Él'sgol'c [54].

Further details on general sequential decision processes may be found in Denardo [31], and Denardo and Mitten [35]. Karp and Held [69] have done some interesting work on finite-state processes and dynamic programming.

† Reprinted with permission from *SIAM Rev.* **9**, 165–177 (1967). Copyright 1967 by Society for Industrial and Applied Mathematics. All rights reserved.

Chapter 9

Conclusion

We have discussed several models of Markovian decision processes and their extended versions. Markovian decision processes have been studied in Chapters 2–5; semi-Markovian decision processes, which generalize to a continuous-time system, have been studied in Chapter 6. The problems have been formulated using the policy iteration algorithm or a linear programming algorithm that implies an optimal strategy. Further, we have shown the relationship between the above two algorithms. The fact of this relationship is of great interest because the linear and dynamic programming approaches are mutually related.

A generalized Markovian decision process in Chapter 7 is an extension of state and action spaces to any Borel sets. Some existence theorems for optimal plans in this model have been studied.

General sequential decision processes have been discussed in Chapter 8. If the contraction (or the N-stage contraction) and the monotonicity assumptions are both satisfied, then the problem for finding optimal strategies is reduced to one for finding fixed points in a complete metric space. Thus, the successive approximation of mathematical analysis, mathematical programming, and policy iteration of Howard type immediately provide optimization schemes.

Several other models besides those described above may be considered. For example, a denumerable Markovian decision process, that is, a Markovian decision process on a denumerable state space, has been studied by Derman [42], Maitra [82], Ross [102], Fisher [58], and others.

The rest of this chapter discusses the history and some applications of Markovian decision processes.

9.1. History

Although the term "Markovian decision process" was used by Bellman [6 and 7, Chapter 7] in 1957, its origin can be found in Shapley's study of the stochastic game [105], as well as in the studies of "games of survival" as examples of multistage games by Bellman, Blackwell, and LaSalle in 1950–1951 (see Bellman [7]). Shapley [105] has stated that if one of two players is a dummy, then the stochastic game is reduced to a dynamic programming problem, that is, a Markovian decision process (see appendix). Further, he has given the equations to obtain v_i for stationary strategies in our terminating process.

Since the appearance of Howard's excellent book [63] in 1960, Markovian decision processes have become well-known and many works have been published. Howard [63] has given the so-called policy iteration algorithms for the discounted and nondiscounted Markovian decision processes and has also solved three famous problems, that is, the taxicab problem, the baseball problem, and the automobile replacement problem using these algorithms. Howard has also considered continuous-time Markov models and has given similar algorithms.

In 1960 Manne [85] and De Ghellinck [27] formulated completely ergodic Markovian decision processes by means of linear programming. D'Epenoux [49] also formulated the discounted problem by linear programming. Markovian decision processes marked an epoch in 1960.

In 1962 Blackwell [14] treated Markovian decision processes as sequential decision processes and demonstrated that there were optimal stationary strategies. In 1962 Derman [36] also showed, using the functional equations of dynamic programming, that there is an optimal stationary strategy and gave a linear programming formulation.

In 1963 Jewell [65, 66] extended these results to semi-Markovian decision processes and gave similar policy iteration algorithms. In 1965 Blackwell [16] published his paper Discounted Dynamic Programming, in which he extended state and action spaces to include Borel sets and gave some existence theorems for optimal plans. In 1965 Brown [19] discussed Markovian decision processes from the viewpoint of dynamic programming. Denumerable Markovian decision processes were also discussed in 1965 by Derman [41] and Maitra [82].

In 1967 Denardo [31] formulated some general sequential decision processes and discussed their properties. Denardo and Mitten [35] and Karp and Held [69] also discussed some general sequential decision processes.

Many other problems have been studied; as of now more than one-hundred papers on Markovian decision processes and their related topics (including applications) have been published. The bibliography includes most, but not all, of these papers.

9.2. Applications

Markov processes are of great use in the study of system sciences, operations research, reliability theory, control theory, and many other fields. Markovian decision processes can, of course, be applied in those fields that have adopted Markov processes as mathematical models.

Manne [85], D'Epenoux [49], and others have applied Markovian decision processes to inventory and production problems. For the maintenance and replacement problems, Derman [38], Derman and Lieberman [45], Klein [72], Kolesar [74, 75], Taylor [110], and others have made similar applications to maintenance and replacement problems. Antelman *et al.* [1] and Derman and Klein [44] have studied the surveillance problem.

Applications to reliability theory have been made by Derman [40] and others. Reich *et al.* [99] have applied the reliability problem of a real-time aircraft simulator. White [117] has discussed sampling inspection plans. Lave [78] has considered quality control. Klein [73] and White [118] have studied the reject allowance problems. Beneš [10, 11] has applied Markovian decision processes to optimal routing in telephone networks. Åström [3], Eaton and Zadeh [53], Aoki [2], Mandl [84], and others have discussed control theory. Relations between an optimal control problem and a Markovian decision process have been discussed by Aoki [2, p. 301] and Mandl [84].

The further development of the theory of Markovian decision processes will demonstrate the wide possibility of its applications to many models.

Appendix

Stochastic Games

The concept of a *stochastic game* was first introduced by Shapley [105] in 1953. In the stochastic game the play proceeds synchronously in a sequence of steps of time from state to state, according to the given probabilities jointly controlled by the two players. The number of states is assumed to be finite. At each step the play is in a certain state i, chosen from a finite set of N states $(i = 1, 2, \ldots, N)$. Let two players denote player I and player II. Player I has finite m_i choices $(k = 1, 2, \ldots, m_i)$ and player II has finite n_i choices $(r = 1, 2, \ldots, n_i)$ in each state i. If the play is in state i, and the two players choose pure strategies $k \, (1 \leqslant k \leqslant m_i)$ and $r \, (1 \leqslant r \leqslant n_i)$, respectively, according to their decisions, then the transition probability of going to state j is given by p_{ij}^{kr} and player I receives the return a_i^{kr} from player II. Player I is to choose a strategy to maximize the total expected return accumulated at the termination of the game, while player II is to choose a strategy to minimize it. By specifying a starting point, we obtain a particular game Γ_i. The term "stochastic game" will refer to the collection $\Gamma = \{\Gamma_i | \, i = 1, 2, \ldots, N\}$.

In addition Shapley has assumed that at each step there is a positive probability

$$1 - \sum_{j=1}^{N} p_{ij}^{kr} = s_i^{kr} > 0 \tag{A.1}$$

of terminating the play. Then this game is called a "terminating stochastic game." In this appendix we illustrate Shapley's results [105] on terminating stochastic games.

We should mention, however, that there is a nonterminating stochastic game in which the play never terminates. For a nonterminating stochastic game we can apply the average return as its value. This game has been studied by Gillette [60].

When we consider a stochastic game in which one of the two players has no choice of strategies (that is, one player is a dummy), the stochastic game reduces to a *Markovian decision process*. Thus stochastic games are of great interest from the viewpoint of Markovian decision processes. Hoffman and Karp [62] have discussed nonterminating stochastic games from this viewpoint. Further works on stochastic games are found in Everett [55], Zachrisson [121], Charnes and Schroeder [22], and others.

A terminating stochastic game specifies a starting state and the data of $N^2 + N$ matrices

$$P_{ij} = (p_{ij}^{kr}|\ k = 1, 2, \ldots, m_i;\quad r = 1, 2, \ldots, n_i) \qquad (A.2)$$

$$A_i = (a_i^{kr}|\ k = 1, 2, \ldots, m_i;\quad r = 1, 2, \ldots, n_i) \qquad (A.3)$$

with i and $j = 1, 2, \ldots, N$, with elements satisfying

$$p_{ij}^{kr} \geqslant 0, \qquad |a_i^{kr}| \leqslant M$$

$$\sum_{j=1}^{N} p_{ij}^{kr} = 1 - s_i^{kr} \leqslant 1 - s < 1 \qquad (A.4)$$

Since any strategy depends on all its history, it is cumbersome to write it. But we can consider a strategy that is independent of each step. That is, the *stationary strategy* can be represented by N-tuples of probability distributions, thus for player I

$$x = (x_1, x_2, \ldots, x_N) \qquad (A.5)$$

where for each i

$$x_i = (x_i^1, x_i^2, \ldots, x_i^{m_i}) \qquad (A.6)$$

which is a probability distribution. We can also write a stationary strategy y for player II. If x_i^k is 0 or 1, then x_i is a *pure strategy*. But x, in general, denotes a *mixed strategy*.

First we consider a matrix game (see, for example, von Neumann and Morgenstern [112]). Given a matrix game B, let val $[B]$ denote its minimax value to player I, and $X[B]$, $Y[B]$ the sets of optimal strategies for players I and II, respectively. If B and C are two matrices of the same size, then it is easily shown that

$$|\text{val } [B] - \text{val } [C]| \leqslant \max_{k,r} |b^{kr} - a^{kr}| \qquad (A.7)$$

Returning to the stochastic game Γ, let $A_i(\alpha)$ define the matrix whose element is

$$a_i^{kr} + \sum_{j=1}^{N} p_{ij}^{kr} \alpha_j \qquad (A.8)$$

where $1 \leqslant k \leqslant m_i$, $1 \leqslant r \leqslant n_i$, and α is the $N \times 1$ column vector whose ith element is α_i. Pick $\alpha^{(0)}$ arbitrary, and define $\alpha^{(t)}$ by the following recursion:

$$\alpha_i^{(t)} = \text{val } [A_i(\alpha^{(t-1)})] \qquad (A.9)$$

for $t = 1, 2, \ldots$. We shall show that the limit of $\alpha^{(t)}$ as $t \to \infty$ exists and is independent of $\alpha^{(0)}$, and that its elements are the values of the infinite games Γ_i.

Consider the transformation T:

$$T\alpha = \beta, \qquad \text{where} \quad \beta_i = \text{val } [A_i(\alpha)] \qquad (A.10)$$

Let define the norm of α as

$$\|\alpha\| = \max_i |\alpha_i| \qquad (A.11)$$

Then we have, from (A.7),

$$\|T\beta - T\alpha\| = \max_i |\text{val}\,[A_i(\beta)] - \text{val}\,[A_i(\alpha)]|$$

$$\leqslant \max_{i,k,r} |\sum_{j=1}^{N} p_{ij}^{kr}\beta_j - \sum_{j=1}^{N} p_{ij}^{kr}\alpha_j|$$

$$\leqslant \max_{i,k,r} |\sum_{j=1}^{N} p_{ij}^{kr}| \max_j |\beta_j - \alpha_j|$$

$$= (1 - s)\|\beta - \alpha\| \qquad (A.12)$$

In particular, $\|T^2\alpha - T\alpha\| \leqslant (1 - s)\|T\alpha - \alpha\|$. Hence the sequence $\alpha^{(0)}$, $T\alpha^{(0)}$, $T^2\alpha^{(0)}$, . . ., is convergent. The limit vector ϕ has the property $\phi = T\phi$, which means that ϕ is a *fixed point* of T. But there is only one such vector, since $\psi = T\psi$ implies

$$\|\psi - \phi\| = \|T\psi - T\phi\| \leqslant (1 - s)\|\psi - \phi\| \qquad (A.13)$$

from (A.12), whence $\|\psi - \phi\| = 0$. Hence ϕ is a unique fixed point of T and is independent of $\alpha^{(0)}$.

To show that ϕ_i is the value of the game Γ_i, we observe that by following an optional strategy of the finite game $\Gamma_i^{(t)}$ for the first steps and by playing arbitrarily thereafter, player I can assure himself an amount within $\varepsilon_t = (1 - s)^t M/s$ of the value of $\Gamma_i^{(t)}$; likewise for player II. Since $\varepsilon_t \to 0$ and the value of $\Gamma_i^{(t)}$ converges to ϕ_i, we conclude that ϕ_i is indeed the value of Γ_i. Thus we have:

THEOREM A.1. *The value of the stochastic game Γ is a unique solution ϕ of the system*

$$\phi_i = \text{val}\,[A_i(\phi)] \qquad \text{for} \quad i = 1, 2, \ldots, N \qquad (A.14)$$

THEOREM A.2. *The stationary strategies x^*, y^*, where $x_i \in X[A_i(\phi)]$ and $y_i \in Y[A_i(\phi)]$ $(i = 1, 2, \ldots, N)$ are optimal for players I and II, respectively, in every game Γ_i belonging to Γ.*

PROOF. Let a finite version of Γ_i be defined by agreeing that on the tth step the play shall stop, with player I receiving the amount $a_i^{kr} + \sum p_{ij}^{kr}\phi_j$ instead of just ϕ_i. Clearly, the stationary strategy x^* assures player I the amount ϕ_i in this finite version. In the original game Γ_i, if player I uses x^*, his expected winnings after t steps will be at least

$$\phi_i - (1 - s)^{t-1} \max_{i,k,r} \sum_{j=1}^{N} p_{ij}^{kr}\phi_j \qquad (A.15)$$

and hence at least

$$\phi_i - (1 - s)^t \max_j \phi_j \tag{A.16}$$

His total winnings are therefore at least

$$\phi_i - (1 - s)^t \max_j \phi_j - (1 - s)^t M/s \tag{A.17}$$

Since this is true for arbitrary large value of t, it follows that x^* is optimal in Γ_i for player I. Similarly, y^* is optimal for player II. ∎

The nonlinearity of the "val" operator often makes it difficult to obtain exact solutions by means of Theorem A.1 and A.2. It therefore becomes desirable to express the payoff directly in terms of stationary strategies. Let $\overline{\Gamma} = \{\overline{\Gamma}_i\}$ denote the collection of games whose pure strategies are the stationary strategies of Γ. Their payoff functions $\mathfrak{H}_i(x, y)$ must satisfy

$$\mathfrak{H}_i(x, y) = x_i A_i y_i + \sum_j x_i P_{ij} y_i \mathfrak{H}_j(x, y) \tag{A.18}$$

for $i = 1, 2, \ldots, N$. This system has a unique solution; indeed, for the linear transformation T_{xy}:

$$T_{xy}\alpha = \beta, \qquad \text{where} \quad \beta_i = x_i A_i y_i + \sum_j x_i P_{ij} y_i \alpha_j \tag{A.19}$$

We have at once

$$\|T_{xy}\beta - T_{xy}\alpha\| = \max_i |\sum_j x_i P_{ij} y_i (\beta_j - \alpha_j) \leqslant (1 - s)\|\beta - \alpha\| \tag{A.20}$$

which corresponds to (A.12). Hence, by Cramer's rule we have

$$\mathfrak{H}_i(x, y) = \frac{\begin{vmatrix} x_1 P_{11} y_1 - 1 & x_1 P_{12} y_1 & \cdots & -x_1 A_1 y_1 & \cdots & x_1 P_{1N} y_1 \\ x_2 P_{21} y_2 & x_2 P_{22} y_2 - 1 & \cdots & -x_2 A_2 y_2 & \cdots & x_2 P_{2N} y_2 \\ \cdots & \cdots & \cdots & \cdots & \cdots & \cdots \\ x_N P_{N1} y_N & x_N P_{N2} y_N & \cdots & -x_N A_N y_N & \cdots & x_N P_{NN} y_N - 1 \end{vmatrix}}{\begin{vmatrix} x_1 P_{11} y_1 - 1 & x_1 P_{12} y_1 & \cdots & x_1 P_{1i} y_1 & \cdots & x_1 P_{1N} y_1 \\ x_2 P_{21} y_2 & x_2 P_{22} y_2 - 1 & \cdots & x_2 P_{2i} y_2 & \cdots & x_2 P_{2N} y_2 \\ \cdots & \cdots & \cdots & \cdots & \cdots & \cdots \\ x_N P_{N1} y_N & x_N P_{N2} y_N & \cdots & x_N P_{Ni} y_N & \cdots & x_N P_{NN} y_N - 1 \end{vmatrix}} \tag{A.21}$$

THEOREM A.3. *The games $\overline{\Gamma}_i$ possess saddle point:*

$$\min_y \max_x \mathfrak{H}_i(x, y) = \max_x \min_y \mathfrak{H}_i(x, y) \tag{A.22}$$

for $i = 1, 2, \ldots, N$. Any stationary strategy that is optimal for all $\overline{\Gamma}_i \in \overline{\Gamma}$ is an optimal pure strategy for all $\overline{\Gamma}_i \in \overline{\Gamma}$, and conversely. The value vectors of Γ and $\overline{\Gamma}$ are the same.

The proof is a simple argument based on Theorem A.2. It should be pointed out that a strategy x may be optimal for one game Γ_i (or $\overline{\Gamma}_i$) and not optimal for other games belonging to Γ (or $\overline{\Gamma}$). This is due to the possibility that Γ might be "disconnected"; however, if none of the p_{ij}^{kr} are zero, this possibility does not arise.

It can be shown that the sets of optimal stationary strategies for Γ are closed, convex polyhedra. A stochastic game with rational coefficients does not necessarily have a rational value. Thus, unlike the minimax theorem for bilinear forms, (A.22) is not valid in an arbitrary ordered field.

Bibliography

1. Antelman, G. R., Russel, C. B., and Savage, I. R., Surveillance problems: two-dimensional with continuous surveillance, *SIAM J. Control* **5**, 245–267 (1967).
2. Aoki, M., "Optimization of Stochastic Systems." Academic Press, New York, 1967.
3. Åström, K. J., Optimal control of Markov processes with incomplete state information, *J. Math. Anal. Appl.* **10**, 174–205 (1965).
4. Barlow, R. E., Applications of semi-Markov processes to counter problems, *in* "Studies in Applied Probability and Management Science" (K. J. Arrow, S. Karlin, and H. Scarf, eds.), Chapter 3. Stanford Univ. Press, Stanford, California, 1962.
5. ———, and Proschan, F., "Mathematical Theory of Reliability." Wiley, New York, 1965.
6. Bellman, R., A Markovian decision process, *J. Math. Mech.* **6**, 679–684 (1957).
7. ———, "Dynamic Programming." Princeton Univ. Press, Princeton, New Jersey, 1957.
8. ———, "Adaptive Control Processes: A Guide Tour." Princeton Univ. Press, Princeton, New Jersey, 1961.
9. ———, and Dreyfus, S., "Applied Dynamic Programming." Princeton Univ. Press, Princeton, New Jersey, 1962.
10. Beneš, V. E., Programming and control problems arising from optimal routing in telephone networks, *SIAM J. Control* **4**, 6–18 (1966).
11. ———, Programming and control problems arising from optimal routing in telephone networks, *Bell System Tech. J.* **45**, 1373–1438 (1966).
12. Bharucha-Reid, A., "Elements of the Theory of Markov Processes and Their Applications." McGraw-Hill, New York, 1960.
13. Blackwell, D., On the functional equation of dynamic programming, *J. Math. Anal. Appl.* **2**, 273–276 (1961).
14. ———, Discrete dynamic programming, *Ann. Math. Statist.* **33**, 719–726 (1962).
15. ———, Memoryless strategies in finite-state dynamic programming, *Ann. Math. Statist.* **35**, 863–865 (1964).
16. ———, Discounted dynamic programming, *Ann. Math. Statist.* **36**, 226–235 (1965).
17. ———, Positive dynamic programming, *Proc. Fifth Berkeley Symp. Math. Statist. Probability* (L. M. Le Cam and J. Neyman, eds.), Vol. I, pp. 415–418. Univ. of California Press, Berkeley, California, 1967.
18. ———, and Ryll-Nardzewski, C., Non-existence of everywhere proper conditional distributions, *Ann. Math. Statist.* **34**, 223–225 (1963).
19. Brown, B. W., On the iterative method of dynamic programming on a finite space discrete time Markov process, *Ann. Math. Statist.* **36**, 1279–1285 (1965).
20. Carton, D. C., Une application de l'algorithm de Howard pour des phénomènes saisonniers, *Proc. 3d Intern. Conf. Operations Res.* pp. 683–691, Oslo, 1963.
21. Charnes, A. and Cooper, W. W., Programming with linear fractional functionals, *Naval Res. Logist. Quart.* **9**, 181–186 (1962).
22. ———, and Schroeder, R. G., On some stochastic tactical antisubmarine games, *Naval Res. Longist. Quart.* **14**, 291–312 (1967).

23. Chung, K. L., "Markov Chains with Stationary Transition Probabilities." Springer, Berlin, 1960.
24. Cox, D. R., "Renewal Theory." Methuen, London, 1962.
25. Dantzig, G. B., "Linear Programming and Extensions." Princeton Univ. Press, Princeton, New Jersey, 1963.
26. De Cani, J. S., A dynamic programming algorithm for embedded Markov chains when the planning horizon is at infinity, *Management Sci.* **10**, 716–733 (1964).
27. De Ghellinck, G. T., Les problème de décisions séquentielles, *Cahiers Centre Études Rech. Opérationnelle* **2**, 161–179 (1960).
28. ———, and Eppen, G. D., Linear programming solutions for separable Markovian decision problems, *Management Sci.* **13**, 371–394 (1967).
29. De Leve, G., "Generalized Markovian Decision Processes: Part I, Model and Method." Mathematical Centre Tracts 3. Mathematisch Centrum, Amsterdam, 1964.
30. ———, "Generalized Markovian Decision Processes: Part II, Probabilistic Background," Mathematical Centre Tracts 4. Mathematisch Centrum, Amsterdam, 1964.
31. Denardo, E. V., Contraction mappings in the theory underlying dynamic programming, *SIAM Rev.* **9**, 165–177 (1967).
32. ———, Separable Markovian decision problems, *Management Sci.* **14**, 451–462 (1968).
33. ———, and Fox, B. L., Multichain Markov revewal programs, *SIAM J. Appl. Math.* **16**, 468–487 (1968).
34. ———, and Miller, B. L., An optimality condition for discrete dynamic programming with no discounting, *Ann. Math. Statist.* **39**, 1220–1227 (1968).
35. ———, and Mitten, L. G., Elements of sequential decision processes, *J. Ind. Eng.* **18**, 106–112 (1967).
36. Derman, C., On sequential decisions and Markov chains, *Management Sci.* **9**, 16–24 (1962).
37. ———, Stable sequential control rules and Markov chains, *J. Math. Anal. Appl.* **6**, 257–265 (1963).
38. ———, Optimal replacement and maintenance under Markovian deterioration with probabilistic bounds on failure, *Management Sci.* **10**, 478–481 (1963).
39. ———, On optimal replacement rules when changes of state are Markovian, *in* "Mathematical Optimization Techniques" (R. Bellman, ed.). Univ. of California Press, Berkeley, California, 1963.
40. ———, On sequential control processes, *Ann. Math. Statist.* **35**, 341–349 (1964).
41. ———, Markovian sequential control processes—denumerable state space, *J. Math. Anal. Appl.* **10**, 295–302 (1965).
42. ———, Denumerable state Markovian decision processes—average cost criterion, *Ann. Math. Statist.* **37**, 1545–1553 (1966).
43. ———, and Klein, M., Some remarks on finite horizon Markovian decision models, *Operations Res.* **13**, 272–278 (1965).
44. ———, and ———, Surveillance of multi-component systems: a stochastic traveling salesman's problem, *Naval Res. Logist. Quart.* **13**, 103–111 (1966).
45. ———, and Lieberman, G. J., A Markovian decision model for a joint replacement and stocking problem, *Management Sci.* **13**, 609–617 (1967).
46. ———, and Strauch, R. E., A note on memoryless rules for controlling sequential control processes, *Ann. Math. Statist.* **37**, 276–278 (1966).

47. Derman, C., and Veinott, Jr., A. F., A solution to a countable system of equations arising in Markovian decision processes, *Ann. Math. Statist.* **38**, 582–584 (1967).

48. D'Epenoux, F., Sur un problème de production et de stockage dans l'Aétoire, *Rev. Franç. Rech. Opérationelle* **14**, 3–16 (1960).

49. ———, A probabilistic production and inventory problem, *Management Sci.* **10**, 98–108 (1963).

50. Doob, J. L., "Stochastic Processes." Wiley, New York, 1953.

51. Dubins, L. E., and Savage, L. J., "How to Gamble If You Must." McGraw-Hill, New York, 1965.

52. Dynkin, E. B., Controlled random sequences, *Theory Probability Appl.* (*USSR*) (*English Transl.*) **10**, 1–14 (1965).

53. Eaton, J. H., and Zadeh, L. A., Optimal pursuit strategies in discrete-state probabilistic systems, *Trans. ASME, Ser. D: J. Basic Eng.* **84**, 23–29 (1962).

54. Él'sgol'c, L. É., "Qualitative Methods in Mathematical Analysis." Am. Math. Soc., Providence, Rhode Island, 1964.

55. Everett, E., Recursive games, *in* "Contributions to the Theory of Games" (M. Dresher, A. W. Tucker, and P. Wolfe, eds.), Vol. III, pp. 47–78. Ann. of Math. Studies, No. 39, Princeton Univ. Press, Princeton, New Jersey, 1957.

56. Feller, W., "An Introduction to Probability Theory and Its Applications," Vol. 1, 2d ed. Wiley, New York, 1957.

57. ———, "An Introduction to Probability Theory and Its Applications," Vol. 2. Wiley, New York, 1966.

58. Fisher, L., On recurrent denumerable decision processes, *Ann. Math. Statist.* **39**, 424–434 (1968).

59. Fox, B., Markov renewal programming by linear fractional programming, *SIAM J. Appl. Math.* **14**, 1418–1432 (1966).

60. Gillette, D., Stochastic games with zero stop probabilities, *in* "Contributions to the Theory of Games" (Dresher, Tucker, and Wolfe, eds.), Vol. III, pp. 179–187, Ann. Math. Studies, No. 39, Princeton Univ. Press, Princeton, New Jersey, 1957.

61. Hatori, H., On Markov chains with rewards, *Kōdai Math. Sem. Rep.* **18**, 184–192 (1966).

62. Hoffman, A. J. and Karp, R. M., On nonterminating stochastic games, *Management Sci.* **12**, 359–370 (1966).

63. Howard, R. A., "Dynamic Programming and Markov Processes." M.I.T. Press, Cambridge, Massachusetts, 1960.

64. ———, Research in semi-Markovian decision structures, *J. Operations Res. Soc. Japan* **6**, 163–199 (1964).

65. Jewell, W. S., Markov-renewal programming: I. Formulation, finite return models, *Operations Res.* **11**, 938–948 (1963).

66. ———, Markov-renewal programming: II. Infinite return models, example, *Operations Res.* **11**, 949–971 (1963).

67. Karlin, S., The structure of dynamic programming models, *Naval Res. Logist. Quart.* **2**, 284–294 (1955).

68. ———, "A First Course in Stochastic Processes." Academic Press, New York, 1966.

69. Karp, R. M., and Held, M., Finite-state processes and dynamic programming, *SIAM J. Appl. Math.* **15**, 693–717 (1967).

70. Kemeny, J. G., and Snell, J. L., "Finite Markov Chains." Van Nostrand, Princeton, New Jersey, 1960.
71. ——, ——, and Knapp, A. W., "Denumerable Markov Chains." Van Nostrand, Princeton, New Jersey, 1966.
72. Klein, M., Inspection-maintenance-replacement schedules under Markovian deterioration, *Management Sci.* **9**, 25–32 (1962).
73. ——, Markovian decision models for reject allowance problems, *Management Sci.* **12**, 349–358 (1966).
74. Kolesar, P., Minimum cost replacement under Markovian deterioration, *Management Sci.* **12**, 694–706 (1966).
75. ——, Randomized replacement rules which maximize the expected cycle length of equipment subject to Markovian deterioration, *Management Sci.* **13**, 867–876 (1967).
76. Kolmogorov, A. N., and Fomin, S. V., "Elements of the Theory of Functions and Functional Analysis." Graylock Press, Rochester, New York, 1957.
77. Krylov, N. V., The construction of an optimal strategy for a finite controlled chain, *Theory Probability Appl.* (*USSR*) (*English Transl.*) **10**, 45–54 (1965).
78. Lave, Jr., R. E., A Markov decision process for economic quality control, *IEEE Trans. Systems Sci. and Cybernetics* **SSC - 2**, 45–54 (1966).
79. Loève, M., "Probability Theory," 3d ed. Van Nostrand, Princeton, New Jersey, 1963.
80. MacQueen, J., A modified dynamic programming method for Markovian decision problems, *J. Math. Anal. Appl.* **14**, 38–43 (1966).
81. ——, A test for suboptimal actions in Markovian decision problems, *Operations Res.* **15**, 559–561 (1967).
82. Maitra, A., Dynamic programming for countable state systems, *Sankhyā Ser. A* **27**, 259–266 (1965).
83. ——, A note on undiscounted dynamic programming, *Ann. Math. Statist.* **37** 1042–1044 (1966).
84. Mandl, P., Analytic methods in the theory of controlled Markov processes, *Trans. Fourth Prague Conf.* pp. 45–53. Academia Praha, Prague, 1967.
85. Manne, A. S., Linear programming and sequential decisions, *Management Sci.* **6**, 259–267 (1960).
86. Martin, J. J., "Bayesian Decision Problems and Markov Chains." Wiley, New York, 1967.
87. Martin-Löf, A., Existence of a stationary control for a Markov chain maximizing the average reward, *Operations Res.* **15**, 866–871 (1967).
88. ——, Optimal control of a continuous-time Markov chain with periodic transition probabilities, *Operations Res.* **15**, 872–881 (1967).
89. Miller, B. L., Finite state continuous time Markovian decision processes with a finite planning horizon, *SIAM J. Control* **6**, 266–280 (1968).
90. ——, Finite state continuous time Markovian decision processes with an infinite planning horizon, *J. Math. Anal. Appl.* **22**, 552–569 (1968).
91. Mitten, L. G., Composition principles for synthesis of optimal multi-stage processes, *Operations Res.* **12**, 610–619 (1964).
92. Norman, D. J., and White, D. J., A method for approximate solutions to stochastic dynamic programming problem using expectations, *Operations Res.* **16**, 296–306 (1968).
93. Ogawara, M., A note on discrete Markovian decision processes, *Bull. Math. Statist.* **11**, 35–42 (1963).

94. Osaki, S., and Mine, H., Linear programming algorithms for semi-Markovian decision processes, *J. Math. Anal. Appl.* **22,** 356–381 (1968).
95. ——, and ——, Some remarks on a Markovian decision problem with an absorbing state, *J. Math. Anal. Appl.* **23,** 327–333 (1968).
96. ——, and ——, Linear programming considerations on Markovian decision processes with no discounting, *J. Math. Anal. Appl.* **26,** 221–232 (1969).
97. Pyke, R., Markov renewal processes: definitions and preliminary properties, *Ann. Math. Statist.* **32,** 1231–1242 (1961).
98. ——, Markov renewal processes with finitely many states, *Ann. Math. Statist.* **32,** 1243–1259 (1961).
99. Reich, W. J., Flannery, W. A., and Miller, D. A., Reliability-maintenability cost trade-off via dynamic and linear programming, *Ann. Reliability and Maintenability* **5,** 310–329 (1966).
100. Riis, J. O., Discounted Markov programming in a periodic process, *Operations Res.* **13,** 920–929 (1965).
101. Romanovski, I. V., Existence of an optimal stationary policy in a Markov decison process, *Theory Probability Appl.* (*USSR*) (*English Transl.*) **10,** 120–122 (1965).
102. Ross, S. M., Non-discounted denumerable Markovian decision models, *Ann. Math. Statist.* **39,** 412–423 (1968).
103. Rykov, V. V., Markov decision processes with finite state and decision spaces, *Theory Probability Appl.* (*USSR*) (*English transl.*) **11,** 302–311 (1966).
104. Shapiro, J. F., Turnpike planning horizons for a Markovian decision model, *Management Sci.* **14,** 292–306 (1968).
105. Shapley, L. S., Stochastic games, *Proc. Natl. Acad. Sci. U.S.A.* **39,** 1095–1100 (1953).
106. Smallwood, R. D., Optimal policy regions for Markov processes with discounting, *Operations Res.* **14,** 658–669 (1966).
107. Smith, W. L., Regenerative stochastic processes, *Proc. Roy. Soc. London, Ser. A* **232,** 6–31 (1955).
108. ——, Renewal theory and its ramifications, *J. Roy. Statist. Soc., Ser. B* **20,** 243–302 (1958).
109. Strauch, R. E., Negative Dynamic Programming, *Ann. Math. Statist.* **37,** 871–890 (1966).
110. Taylor, III, H. M., Markovian sequential replacement processes, *Ann. Math. Statist.* **36,** 1677–1694 (1965).
111. Veinott, Jr., A. F., On the finding optimal policies in discrete dynamic programming with no discounting, *Ann. Math. Statist.* **37,** 1284–1294 (1966).
112. von Neumann, J., and Morgenstern, O., "Theory of Games and Economic Behaviour," 3d ed. Princeton Univ. Press, Princeton, New Jersey, 1953.
113. Wagner, H. M., On the optimality of pure strategies, *Management Sci.* **6,** 268–269 (1960).
114. Wedekind, H., Primal und Dual-Algorithmen zur Optimierung von Markov-Prozessen, *Unternehmensforschung* **8,** 128–135 (1964).
115. Weitzman, M., On choosing an optimal technology, *Management Sci.* **13,** 413–428 (1967).
116. White, D. J., Dynamic programming, Markov chains, and the method of successive approximations, *J. Math. Anal. Appl.* **6,** 373–376 (1963).

117. White, L. S., Markovian decision models for the evaluation of a large class of continuous sampling inspection plans, *Ann. Math. Statist.* **36,** 1408–1420 (1965).
118. ———, Bayes Markovian decision models for a multiperiod reject allowance problem, *Operations Res.* **15,** 857–865 (1967).
119. Widder, D. V., "The Laplace Transforms." Princeton Univ. Press, Princeton, New Jersey, 1946.
120. Wolfe, P., and Dantzig, G. B., Linear programming in a Markov chain, *Operations Res.* **10,** 702–710 (1962).
121. Zachrisson, L. E., Markov games, *in* "Advances in Game Theory" (M. Dresher, L. S. Shapley, and A. W. Tucker, eds.), pp. 211–253, Ann. of Math. Studies, No. 52. Princeton Univ. Press, Princeton, New Jersey, 1964.
122. Fox, B., Existence of stationary optimal policies for some Markov renewal programs, *SIAM Rev.* **9,** 573–576 (1967).

AUTHOR INDEX

Numbers in parentheses indicate the numbers of the references when these are cited in the text without the names of the authors.

Numbers set in *italics* designate the page numbers on which the complete literature citation is given.

Names listed with page numbers only in *italics* refer to the authors of publications listed in the Bibliography but not cited in the text.

SUBJECT INDEX